The Development of Renewable Energy Sources and its Significance for the Environment

Almas Heshmati • Shahrouz Abolhosseini
Jörn Altmann

The Development of Renewable Energy Sources and its Significance for the Environment

 Springer

Almas Heshmati
Sogang University
Seoul, Korea, Republic of (South Korea)

Jörn Altmann
College of Engineering,
Seoul National University
Seoul, Korea, Republic of (South Korea)

Shahrouz Abolhosseini
College of Engineering,
Seoul National University
Seoul, Korea, Republic of (South Korea)

ISBN 978-981-10-1273-0   ISBN 978-981-287-462-7   (eBook)
DOI 10.1007/978-981-287-462-7

Springer Singapore Heidelberg New York Dordrecht London
© Springer Science+Business Media Singapore 2015
Softcover reprint of the hardcover 1st edition 2015

Printed on acid-free paper

Springer Science+Business Media Singapore Pte Ltd. is part of Springer Science+Business Media (www.springer.com)

Preface

The ongoing concerns about climate change have made renewable energy sources an important topic of research. Several scholars have applied different methodologies for examining the relationship between energy consumption, environment and economic growth of individual countries and groups of countries, in order to understand the effects of energy policies. In particular, previous studies have analyzed carbon dioxide emission savings made through the use of renewable energy from an individual source or in combination with traditional sources of energy by applying life-cycle analysis methods. This research has shown that after a certain period, economic growth leads to the promotion of the environmental quality. However, econometric critiques have opposed the results of these studies. Moreover, the effectiveness of governance-related parameters has been neglected in these studies. In this research, among others we analyze the impact of renewable energy development on carbon dioxide emission reduction.

In this volume, a number of issues are discussed that play a crucial role in enhancing the deployment of renewable energy, namely, the energy–environment relationship, alternative renewable energy production technologies, regulation frameworks for renewable energy sources, financing renewable energy development, and the market design for trading commoditized electricity generated by small renewable energy sources. Local power generation, which is the basis of renewable energy production, encourages the production of renewable energy resources, decreases transmission loss, increases saving energy, and enhances energy efficiency. Therefore, the integration of local, renewable energy sources and smart grids through local marketplaces that trade renewable energy in small units is a promising solution.

There are several points making this book unique compared to others. It analyzes important aspects of renewable energy development and its challenges. A model is developed to evaluate the effectiveness of renewable energy development, technological innovation, market for trade, and market regulations with respect to carbon dioxide emission reduction. For this purpose, a panel data model is applied to data from the EU-15 countries between 1995 and 2010. The effects of renewable energy on carbon emission reduction in EU-15 is investigated. The findings show that the effects of climate change can be mitigated by governance-related parameters in addition to regulations, economic incentives, and technology development measures. It proposes a marketplace for trading renewable energy sources and provides suitable and evidence-based policy recommendations to promote renewable energies to substitute fossil fuels.

The subject of this work is development of renewable energy sources and their significance for the environment. A number of issues of particular interest to the readers are raised. We present the development of different renewable sources in recent decades and forecasts for future illustrated in figures and tables. Some regression analysis is also used for establishing relationship between emission and use of renewable energies. The key features of this work is its deep review and analysis of technologies, finances, environment and trade markets for renewable energy sources. It provides an up-to-date review of the literature considering production and consumption of renewable energy sources at country, regional and global levels.

Deployment of renewable energy and technological innovations can be used to reduce carbon emissions. Tariffs, finances, tax policies, and energy efficiency are used by governments to develop renewable energy. State research and development support, innovation, finances, and regulations have impacted the market for renewable energies. The effects of different technology, regulations and financial support factors on emission reductions are estimated. The structure of a marketplace for renewable energy sources is proposed and the requirements for the marketplace to function are outlined. Suitable policy recommendations are provided to enhance the efficient operation of market for renewable energies. Researchers, professionals, decision makers, environmentalists, non-governmental organizations, graduate students, postgraduate students, and public and private utilities will benefit from reading this research.

Seoul, Korea, Republic of (South Korea)   Almas Heshmati
                                                                        Shahrouz Abolhosseini
                                                                              Jörn Altmann

Contents

Biographies of the Authors

Almas Heshmati Almas Heshmati is Professor of Economics at Jönköping University and Sogang University. He held similar positions at the Korea University, Seoul National University, University of Kurdistan Hawler, RATIO Institute (Sweden), and the MTT Agrifood Research (Finland). He was Research Fellow at the World Institute for Development Economics Research (WIDER), the United Nations University during 2001–2004. From 1998 to 2001, he was Associate Professor of Economics at the Stockholm School of Economics. He has a Ph.D. degree from the University of Gothenburg (1994), where he held a Senior Researcher position until 1998. His research interests include applied microeconomics, globalization, development strategy, efficiency, productivity, and growth with application to manufacturing and services. In addition to more than 100 scientific journal articles, he has published books on the EU Lisbon Process, global inequality, East Asian manufacturing, the Chinese economy, technology transfer, information technology, water resources, landmines, power generation, development economics, economic growth, and world values.

Shahrouz Abolhosseini Shahrouz Abolhosseini has a Ph.D. degree in Technology Management (2014) from Seoul National University. Prior to his doctoral studies, he served as a crude oil marketing expert (2004–2007) and petroleum products marketing expert (2007–2010) at the National Iranian Oil Company. Previously he worked as an oil and gas market analyst (2001) and project financial coordinator (2001–2004) at Petropars Limited Company. He held a number of other research-related positions during the period 1995–2000 at the Institute for International Energy Studies, Economic and Finance Affairs Ministry, and Parliament Research Center. His research interests include energy economics and policy, and renewable and non-renewable energy sources.

Jörn Altmann Jörn Altmann is Professor for Technology Management, Economics, and Policy at the College of Engineering of Seoul National University. Prior to this, he taught at the University of California at Berkeley, worked as a Senior Scientist at Hewlett-Packard Labs, and has been a postdoc at EECS and ICSI of UC Berkeley. During that time, he worked on international research projects about pricing of Internet services. Dr. Altmann received his B.Sc., his M.Sc. (1993), and his Ph.D. (1996) from the University of Erlangen-Nürnberg, Germany. Dr. Altmann's current research centres on the economics of Internet services and Internet infrastructures that integrate economic models. In particular, he focuses on the analyses of network topologies, networked businesses, and resource allocation schemes. On these topics of research, he has publications in major conferences and journals, serves on editorial bodies of journals (*Electronic Markets, Electronic Commerce Research Journal*), is involved in many conference program committees, and has been an invited speaker to several conferences. He also served on several European, US American, and other national panels for evaluating research proposals on networks and emerging technologies. Currently, he chairs the International Conference on Economics on Grids, Clouds, Systems, and Services, GECON.

List of Abbreviations

AIC	Akaike information criterion
APEC	Asia-Pacific Economic Cooperation
ASEAN	Association of Southeast Asian Nations
BIC	Bayesian information criterion
BRICS	Brazil, Russia, India, China, and South Africa
CCGT	Combined cycle gas turbine
CCHP	Combined cooling heat and power
CCMT	Carbon-change-mitigation technology
CDM	Clean development mechanism
CHP	Combined heat and power
CSP	Concentrated solar power plant
CVPP	Commercial virtual power plant
DER	Distributed energy resources
DG	Distributed generation
EEX	European Energy Exchange
EIA	Energy Information Administration
EKC	Environmental Kuznets curve
EMS	Environmental management system
EPBT	Energy payback time
ERGO	Electric recharging grid operator
EROI	Energy return on investment
ETS	Emission trading system
EV	Electric vehicle
EXAA	Energy Exchange Austria
FDI	Foreign direct investment
FE	Fixed-effect
FGLS	Feasible generalized least square
FIT	Feed-in tariff
GDP	Gross domestic product
GHG	Greenhouse gas
GLS	Generalized least square

GNP	Gross national product
GSHP	Ground source heat pump
Gt	Gigatonne
Gtoe	Gigatonnes of oil equivalent
GW	Gigawatt
ICT	Information and communication technology
IEA	International Energy Agency
IPCC	Intergovernmental Panel on Climate Change
IRR	Internal rate of return
JI	Joint implementation
kWh	Kilowatt hour
LCA	Life cycle analysis
MENA	Middle East and North Africa
MMT	Million tones
Mtoe	Million tones of oil equivalent
MW	Megawatt
NPV	Net present value
OECD	Organisation for Economic Co-operation and Development
PHES	Pumped hydro energy storage
PHEV	Plug-in hybrid electric vehicle
PHS	Pumped hydro storage
PTC	Production tax credit
PURPA	Public Utility Regulatory Policies Act of 1978 (US)
PV	Photovoltaic
RE	Renewable energy
RE	Random effect
RES	Renewable energy sources
RET	Renewable energy technology
RPS	Renewable portfolio standard
RPT	Renewable energy premium tariff
SAPV	Stand-alone solar photovoltaic
SHPP	Small-hydro power plant
SOFC	Solid oxide fuel cell
TJ	Terajoule
TOU	Time-of-use
TWh	Terawatt hour
V2G	Vehicle to grid
VAR	Vector autoregression
VPP	Virtual power plant

List of Figures

List of Tables

Chapter 1
Introduction

1.1 Background

Industry's electricity consumption will comprise an increasing share of the global energy demand during the next two decades. It is expected that the growth rate of electricity consumption will be more than that of the consumption of other sources of energy (e.g., liquid fuels, natural gas, and coal) (IEA 2012). The increasing prices of fossil fuels such as crude oil and the increasing concerns about the environmental consequences of greenhouse gas emissions have renewed the interest in the development of alternative energy resources. In particular, the Fukushima Daiichi accident was a turning point in the call for alternative energy sources. Renewable energy is now considered a more desirable source of fuel than nuclear power plants because of the absence of fatal risks.

Considering that carbon dioxide is the major greenhouse gas (GHG), there is a global concern about reducing carbon dioxide emissions. Different policies can be applied in this regard (e.g., enhancing renewable energy deployment and encouraging technological innovations). In addition, supporting mechanisms (e.g., feed-in tariffs, renewable portfolio standards, and tax policies) can be employed by governments to increase renewable energy generation and achieve energy efficiency. Many countries have started installing facilities for power generation that can use renewable energy sources. However, the share of a renewable energy supply differs by region and country. Europe is considered at the forefront of using renewable energy technologies.

The research literature on the relationship between energy consumption and economic growth is extensive. Many researchers have studied the effectiveness of conservative energy policies on economic activities. Some researchers (Fthenakis et al. 2008; Crawford 2009; Frick et al. 2010) have measured the amount of carbon saving by using the life-cycle analysis method. Other researchers have analyzed carbon emission saving by enhancing energy efficiency through cogeneration and advanced technology (Shipley et al. 2008; Kiviluoma and Meibom 2010;

© Springer Science+Business Media Singapore 2015
A. Heshmati et al., *The Development of Renewable Energy Sources and its Significance for the Environment*, DOI 10.1007/978-981-287-462-7_1

Wille-Haussmann et al. 2010). However, no previous study has measured the amount of carbon emission reduction and the interaction effects of different policy tools that support mechanisms to enhance renewable energy sources (generation and consumption), technological innovation, and market regulation.

The methodology used by early researchers to investigate the relationship between emissions and gross domestic product (GDP) per capita is not appropriate. Some researchers such as Stern (2004), Müller-Fürstenberger and Wagner (2007), and Wagner (2008) have cast doubt on the existence of an inverted U-shaped curve showing the relation between carbon emissions and GDP per capita. They argued that the results were obtained by commonly used estimation methods that have serious problems. For instance, the issues of causality and its direction are well established.

Furthermore, a study (Dasgupta et al. 2004) pointed out that this relationship is not as rigid as proposed as poor countries were mistakenly assumed to not have strong governance. The role of GDP growth in CO_2 emission reduction could be reduced by the regulations applied by the governments of such countries. In addition, other parameters such as technological innovation and environmental tax could play an important role in emission reduction. The direct impact of each parameter might change when it is affected by the impact of interactions between different variables.

1.2   The Objective

This research aims to analyze the effects of power generated by renewable energy sources, renewable energy production technology, energy efficiency, and market regulation on carbon emissions. These parameters have direct and indirect effects on carbon emission reduction. For example, environmental tax could reduce carbon emissions directly by decreasing fossil fuel consumption or stimulating energy savings through technological innovation. In addition, renewable energy sources could affect both economic growth and the environment. After analyzing renewable energy consumption, production technology, market regulation, and their relations in detail, we devised a model to measure the extent of their effectiveness and the result of interactions between these parameters. Based on these results, we proposed the structure of a marketplace for renewable energy sources and outlined the requirements for this market to function effectively.

As Europe is considered to be at the forefront of renewable energy deployment, this study selected the EU-15 countries[1] to examine the effects of renewable energy generation on carbon dioxide emission reduction. We examine the long-term effects

---

[1]The EU-15 comprised the following 15 countries: Austria, Belgium, Denmark, Finland, France, Germany, Greece, Ireland, Italy, Luxembourg, The Netherlands, Portugal, Spain, Sweden, and United Kingdom.

of related policies on carbon dioxide emissions of individual countries and the group of EU-15 countries. We compare the effect of each variable over time and across countries. Three hypotheses are posed:

1. The power generated by renewable energy sources in the EU-15 has been able to affect carbon dioxide emission through the displacement of traditional capacity fueled by fossil fuels. Moreover, we expect a negative elasticity for renewable energy sources regarding carbon dioxide emission.
2. Technological advances are able to decrease carbon dioxide emissions by decreasing the costs of renewable energy sources and enhancing energy efficiency. Therefore, we expect a negative relation between technological innovation and carbon dioxide emission.
3. Environmental taxes applied by governments have a direct negative relation with carbon dioxide emissions. The size of this parameter could indicate its importance in comparing renewable energy development and technological innovation. We expect negative elasticity for environmental tax.

We review the relevant literature on the effectiveness of renewable energy development, production technology, and market regulation on reducing carbon dioxide emissions. Based on this, we derive appropriate variables for measuring their impacts on carbon dioxide emission reduction. The effectiveness of technological innovation will be determined by examining patent applications that adopt climate change mitigation and information and communications technology (ICT) patent applications. We apply the panel data method to develop our model in the form of the translog function to investigate the interaction effects of different parameters. After estimation of the model, the elasticity of carbon dioxide emission in relation to GDP, renewable energy generation, energy patent applications, ICT patents, and environmental taxation trends is calculated. In economics, elasticity is the measurement of how change in one variable affects another variable, assuming that all other variables are kept constant. We use this term to measure the effectiveness of the aforementioned variables on carbon dioxide emission.

Our results help identify the variables that have a greater impact on carbon dioxide emission reduction. In addition, the results indicate that policymakers should apply the policies that are the most effective in achieving targets.

The contribution of this research lies in defining three variables (i.e., renewable energy generation, technological innovation, and environmental tax) with regard to the Environmental Kuznets Curve and analyzing their effects on carbon dioxide emission per capita. For the analysis, we employ the number of patents and the amount of environmental taxes for measuring technology and market regulation impacts instead of research and development (R&D) expenditures, which were used by previous research. We also calculate the elasticity of carbon dioxide emissions per capita over time for each EU-15 country and for all EU-15 countries jointly. We also apply an estimation methodology to overcome the econometric issues neglected by the early researchers. Many researchers have estimated fixed-effect models without applying regression diagnostic tests (Cropper and Griffiths 1994; Shafik 1994; Horvath 1997; Moomaw and Unruh 1997; and Suri and Chapman 1998). Our estimation method differs from most studies in its use of feasible generalized least squares (FGLS) to correct heteroscedasticity and autocorrelation. The FGLS method is appropriate for this, as demonstrated by Stern (2002), Aldy (2005), and Luzzati and Orsini (2009) in their research.

1.3   The Outline

In the next chapter, we focus on the current situation of renewable energy consumption and the global outlook. We also review the roles of economic growth, energy security, and carbon dioxide emission reduction as the main drivers in the development of renewable energy. Chapter 3 provides a review of the literature on renewable energy supply technologies and energy efficiency technologies. While we analyze different regulations for increasing the renewable energy deployments in Chap. 4, we focus on financial supporting mechanisms and cross-national incentive policies for enhancing renewable energy deployment in Chap. 5. Based on these results, we propose the requirements and structure of a marketplace for trading small units of renewable energy in Chap. 6. In Chap. 7, we describe our model for evaluating the impact of renewable energy generation, economic growth, technological innovation, and environmental tax on carbon dioxide emission reduction in EU. The results of this model, which are discussed in Chap. 8, could be used by governments to make effective policies to achieve their targets of carbon dioxide emission reduction and climate change mitigation.

References

Aldy JE (2005) An environmental Kuznets curve analysis of US state-level carbon dioxide emissions. J Environ Develop 14(1):48–72

Crawford R (2009) Life cycle energy and greenhouse emissions analysis of wind turbines and the effect of size on energy yield. Renew Sust Energ Rev 13(9):2653–2660

Cropper M, Griffiths C (1994) The interaction of population growth and environmental quality. Am Econ Rev 84(2):250–254

Dasgupta S, Hamilton K, Pandey K, Wheeler D (2004) Air pollution during growth: accounting for governance and vulnerability. World Bank Policy Research Working Paper 3383

Frick S, Kaltschmitt M, Schröder G (2010) Life cycle assessment of geothermal binary power plants using enhanced low-temperature reservoirs. Energy 35(5):2281–2294

Fthenakis VM, Kim HC, Alsema E (2008) Emissions from photovoltaic life cycles. Environ Sci Technol 42(6):2168–2174

Horvath RJ (1997) Energy consumption and the environmental Kuznets curve debate. Department of Geography, University of Sydney, Sydney

IEA (2012) Energy technology perspectives 2012. OECD Publishing, Paris

Kiviluoma J, Meibom P (2010) Influence of wind power, plug-in electric vehicles, and heat storages on power system investments. Energy 35(3):1244–1255

Luzzati T, Orsini M (2009) Investigating the energy-environmental Kuznets curve. Energy 34(3):291–300. http://dx.doi.org/10.1016/j.energy.2008.07.006

Moomaw WR, Unruh GC (1997) Are environmental Kuznets curves misleading us? The case of CO2 emissions. Environ Dev Econ 2:451–463

Müller-Fürstenberger G, Wagner M (2007) Exploring the environmental Kuznets hypothesis: theoretical and econometric problems. Ecol Econ 62(3):648–660

Shafik N (1994) Economic development and environmental quality: an econometric analysis. Oxford Econ Pap 46:757–773

Shipley MA, Hampson A, Hedman MB, Garland PW, Bautista P (2008) Combined heat and power: effective energy solutions for a sustainable future. Oak Ridge National Laboratory (ORNL), Oak Ridge

Stern DI (2002) Explaining changes in global sulfur emissions: an econometric decomposition approach. Ecol Econ 42(1–2):201–220. http://dx.doi.org/10.1016/S0921-8009(02)00050-2

Stern DI (2004) The rise and fall of the environmental Kuznets curve. World Dev 32(8):1419–1439

Suri V, Chapman D (1998) Economic growth, trade and energy: implications for the environmental Kuznets curve. Ecol Econ 25(2):195–208

Wagner M (2008) The carbon Kuznets curve: a cloudy picture emitted by bad econometrics? Resour Energy Econ 30(3):388–408

Wille-Haussmann B, Erge T, Wittwer C (2010) Decentralised optimisation of cogeneration in virtual power plants. Sol Energy 84(4):604–611

Chapter 2
The Energy and Environment Relationship

2.1 Introduction

An expanding body of research shows that there is a strong relation between climate change and the carbon dioxide (CO_2) emissions that are produced through energy production and consumption.

Carbon dioxide emission pollutants are primarily produced by combustion of fossil fuels. According to the International Energy Agency (IEA) estimates, the share of energy production and consumption in carbon dioxide emissions was 81.6 % in 2010 (IEA 2012b). Therefore, energy consumption is the main cause of climate change. According to the *International Energy Outlook 2011* (IEO 2011), the global energy-related carbon dioxide emissions have risen from 30.2 billion metric tons in 2008 to 35.2 billion metric tons in 2020 and will rise to 43.2 billion metric tons in 2035 (Conti and Holtberg 2011). Developing non-OECD (Organization for Economic Co-operation and Development) countries that continue to be heavily dependent on fossil fuel consumption account for much of this growth. These countries need to meet their continuously rising energy demand. Moreover, fossil fuels are subsidized in many countries.

Government policies have played a crucial role in the recent growth in renewable energy sources, especially in the electric power sector. Reducing carbon dioxide emission and local pollutants constitutes a core part of environmental concerns. More than 70 countries are expected to implement policies for deploying renewable energy technologies in the power sector by 2017 (IEA 2012c). Among other objectives, these policies need to achieve an increase in power generation through renewable energy sources so that the unit cost decreases to the level of other energy sources.

Henrik Lund (2010) defines renewable energy as "energy that is produced by natural resources—such as sunlight, wind, rain, waves, tides, and geothermal heat—

© Springer Science+Business Media Singapore 2015
A. Heshmati et al., *The Development of Renewable Energy Sources and its Significance for the Environment*, DOI 10.1007/978-981-287-462-7_2

that are naturally replenished within a time span of a few years." According to his view, all technologies that are able to convert natural resources (e.g., solar) to any kind of energy could help in the generation of renewable energy.

2.2   The General Trend of Energy Consumption

Energy consumption depends on different factors such as economic progress, population, energy prices, weather, and technology. Global consumption of primary energy in 2011 was 12.2 Gtoe (BP 2012). The consumption of crude oil, natural gas, and coal was 4.1, 2.9, and 3.7 Gtoe respectively. The USA, China, and Japan have been the major oil consumers at 833.6, 461.8, and 201.4 Mtoe respectively. While the USA, Russia, and Iran are the biggest consumers of natural gas at 626, 382.1, and 138 Mtoe, China is the biggest consumer in the coal market at 1.8 Gtoe followed by USA and India at 501.9 and 295.6 Mtoe.

According to the *BP Statistical Review of World Energy* (BP 2012), the average primary energy consumption has been 2,306.7 Mtoe during 2001–2010 compared with 2,140.5 Mtoe in 1991–2000, which shows a growth rate of 7.8 % per year. On the other hand, the average carbon dioxide emission was 6,315.9 Mtoe in 2001–10 as against 5,882.7 Mtoe in 1991–2000, showing a growth rate of 7.4 % per year. About 87 % of primary energy consumption in 2010 was from fossil fuels, while the share of nuclear energy, hydroelectricity, and renewable energy was 5.2, 6.5, and 1.4 % respectively. Compared with the primary energy consumption in 2011, the share of fossil fuels has barely changed, but the share of nuclear energy and hydroelectricity has decreased to 4.9 and 6.4 respectively, while the share of renewable energy has gone up to 1.6 %. Table 2.1 shows the global primary energy consumption by types of fuel.

Table 2.1   Global primary energy consumption, end of 2011 (Mtoe)

Region	Oil	Gas	Coal	Nuclear	Hydro	Renewable	Total
N. America	1,026.4	782.4	533.7	211.9	167.6	51.4	2,773.3
S. & C. America	289.1	139.1	29.8	4.9	168.2	11.3	642.5
Europe & Eurasia	898.2	991.0	499.2	271.5	179.1	84.3	2,923.4
Middle East	371.0	362.8	8.7	NA	5.0	0.1	747.5
Africa	158.3	98.8	99.8	2.9	23.5	1.3	384.5
Asia Pacific	1,316.1	531.5	2,553.2	108.0	248.1	46.4	4,803.3
Total World	**4,059.1**	**2,905.6**	**3,724.3**	**599.3**	**791.5**	**194.8**	**12,274.6**
OECD	2,092.0	1,386.1	1,098.6	487.8	315.1	148.0	5,527.7
Non-OECD	1,967.0	1,519.5	2,625.7	111.5	476.4	46.8	6,746.9
EU	645.9	403.1	285.9	205.3	69.6	80.9	1,690.7
FSU	190.6	539.6	169.8	60.2	54.6	0.4	1,015.1

Source: *BP Statistical Review of World Energy, 2012*

2.2.1 Fossil Fuels

According to the *BP Statistical Review of World Energy* (BP 2012), at the end of 2011, 48.1 % of the proven oil reserves were located in the Middle East. As we see in Table 2.1, Europe and Eurasia have 8.5 % of the reserves, of which a majority is located in the Russian Federation (5.3 %) and Kazakhstan (1.8 %). Africa has 8 % of the global oil reserves, mostly in Libya (2.9 %) and Nigeria (2.3 %). In South America, the proven oil reserves are mostly located in Venezuela (91 % of the regional reserves and 17.9 % of the global reserves). North America has 13.2 % of oil reserves, most of which belongs to Canada (80.6 % of regional reserves and 10.6 % of total global reserves). This means that 87 % of proven oil reserves in the American continent are located in Venezuela and Canada. Natural gas reserves are more concentrated geographically than crude oil because 38.4 % of the reserves are located in the Middle East and 37.8 % can be found in Europe and Eurasia. Russia, Iran, and Qatar have almost half the global natural gas reserves. If we take a look at coal reserves, we will find that around 60 % of the global coal reserves are located in the USA, Russia, and China.

In terms of consumption, the share of the Middle East in global oil consumption is 9.1 % (BP 2012). The share of Europe and Eurasia is 22.1 % of the global oil consumption, which is less than the total oil consumption for China, India, Japan, and Korea. Africa has the least share of consumption, with 3.9 %, while North America has a share of 25.3 %. The USA has a share of 20.5 %, almost as much as Europe. This level of North American consumption is more than all countries in the European Union together. The Asia Pacific region has the biggest share in oil consumption, with 32.4 %. China is the second biggest consumer in the world (11.4 %), but its consumption is almost half of the USA. It would be interesting if we compare these numbers with the oil reserves in the USA (2 %) and China (10 %). China is the biggest energy consumer in the world followed by the USA, but the composition of fuel sources is different in these countries. Oil is the main source of energy consumption in the USA, while coal is the most important source of energy in China. Coal consumption in China was 1839.4 Mtoe in 2011, while that of oil and gas was 461.8 and 117.6 Mtoe respectively. Global oil consumption in 2011 has increased by only 0.7 % compared with 2010 because of the economic recession in the major oil consumer countries. Although the oil consumption growth rate is negative in 2011 for OECD countries (−1.2 %), it has been calculated as 2.8 and 5.7 % for the non-OECD and former Soviet Union (FSU) countries respectively. If we compare the growth rate of crude oil consumption and production, we will find that the former (0.7 %) is less than the latter (1.3 %) globally. But this varies across regions. Table 2.2 shows the growth rate of fossil fuel consumption in different regions around the world.

Based on these figures, we find that the consumption growth rate in Asia Pacific is much stronger than other regions. Although the oil consumption growth rate is negative in some regions like Africa and Europe, the rate of decrease in production is much higher than consumption. This means that there is a shortage of supply

Table 2.2 Fossil Fuels production and consumption growth rate during 2010–2011

	Oil		Gas		Coal	
	Prod.	Con.	Prod.	Con.	Prod.	Con.
North America	3.0	−1.4	5.5	3.2	1.2	−4.6
S. & Cent. America	1.3	2.9	3.0	2.9	13.3	5.7
Europe & Eurasia	−1.8	−0.6	0.9	−2.1	4.5	3.3
Middle East	9.3	1.8	11.4	6.9	–	2.1
Africa	−12.8	−1.4	−5.1	2.7	0.3	1.7
Asia Pacific	−2.0	2.7	−0.9	5.9	7.8	8.4
World	1.3	0.7	3.1	2.2	6.1	5.4

Source: *BP Statistical Review of World Energy, 2012*

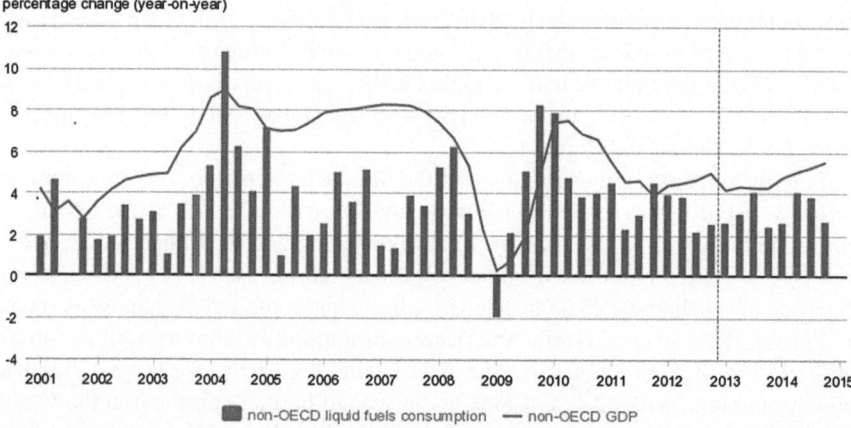

Fig. 2.1 Non-OECD quarterly liquid fuels consumption and GDP (Source: U.S. Energy Information Administration, IHS Global Insight. Updated: Monthly|Last Updated: 2/12/2013)

in these countries. The Middle East is the only region in the world that has a big difference between the production and consumption growth rates for oil and gas.

As mentioned earlier, population growth and expanding economies are the main drivers for increasing energy consumption. According to an IEA report, "world population is projected to grow from an estimated 6.8 billion in 2010 to 8.6 billion in 2035 or by some 1.7 billion new energy consumers" (IEA 2012c). According to *IEA Outlook*, global GDP will increase at a rate of 3.5 % during 2010–2035. It predicts that economic growth in the non-OECD countries will be much more than the OECD countries. The other parameter for driving energy consumption is price. Of course, its direction may not be the same in different countries. Figures 2.1 and 2.2 show the strong impact of GDP on energy consumption in non-OECD countries and price effect in OECD countries.

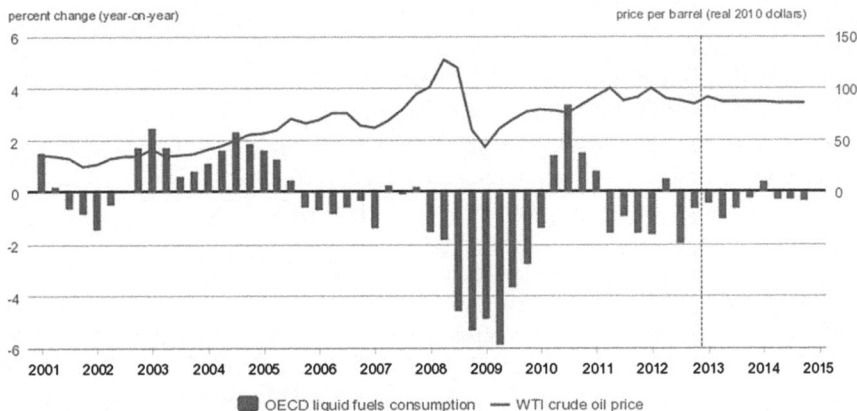

Fig. 2.2 OECD quarterly liquid fuels consumption growth rate and WTI crude oil price development (Source: U.S. Energy Information Administration, Thomson Reuters. Updated: Monthly|Last Updated: 2/12/2013)

Oil consumption in the non-OECD has increased very fast in recent years. The growth rate of oil consumption in these countries in 2010 was more than 40 % compared with its level in 2000, while oil consumption in the OECD countries has decreased in this period. The largest growth in oil consumption has taken place in China, India, and Saudi Arabia (Conti and Holtberg 2011). Increasing oil demand indicates economic advancement in non-OECD countries. Commercial and individual transportation, manufacturing processes, and fuel for power generation in some countries require a huge amount of oil. On the other hand, the population of many non-OECD countries has increased, which supports this trend. Figure 2.1 shows that oil consumption decreased only in the fourth quarter of 2008 and the first quarter of 2009. Oil prices increased sharply in this period, but economic growth in these countries was influenced less than the corresponding for OECD countries (EIA 2013).

OECD countries consume more oil than non-OECD countries, but as Fig. 2.2 shows, the former have a lower oil consumption growth compared with the latter. Oil consumption in OECD countries has decreased from 2,217.3 Mtoe in 2000 to 2,092 Mtoe in 2011, while it increased in non-OECD countries from 1,354.5 to 1,967 Mtoe in the same period (BP 2012).

Due to the different economic structures in OECD and non-OECD countries, oil consumption follows different patterns in these countries. Many developing countries are using energy incentive technologies; also, they do not consider fuel efficiency in economic activities to the same extent as developed countries. In OECD countries, a higher rate of fuel taxes and even carbon tax is imposed on crude oil and petroleum products. Also, they try to improve fuel efficiency economy through policies and new technologies. It is interesting to note that there are some structural differences in energy consumption within economic sections in these countries. Vehicle ownership per capita in developed countries is higher than that of developing countries. There are many households in OECD countries that have more than one car, but this rate is lower for non-OECD countries. Therefore, the transportation sector usually has a bigger share of oil consumption in the former compared with the latter. Furthermore, the size of the service sector in developed countries is larger than developing countries; also, the effect of economic growth on oil consumption is not the same in these countries (EIA 2013).

According to *International Energy Outlook* (*IEO*), world energy consumption will increase by 53 % between 2008 and 2035 (Conti and Holtberg 2011). Although worldwide energy consumption has been limited by the global recession, world energy demand has started to increase with economies recovering from the recession. Economic recovery varies among OECD countries. For example, economic recession has officially ended in the USA, but the recovery is not as strong as those from past recessions. Also, there is a time lag for economic recovery in Europe. It is forecast that world energy demand will increase to a large extent as a result of economic growth in developing countries. Among these countries, China and India were least affected by the recession and continue to lead world economic growth and energy demand (Conti and Holtberg 2011)

The world is dependent on fossil fuels to generate electricity power, which is used for different purposes including industrial, agricultural, commercial, and residential consumption. Currently, the growth rate of energy consumption is 2.5 % per annum (BP 2012). Mason (2007) has mentioned that if energy consumption continues to grow at the rate of 2 %, then it will double in 35 years, which increases the urgency of concerns regarding energy sources. In this context, many scholars have tried to estimate the size of global fossil fuel reserves and the time it will take to diminish these reserves. Salameh (2003) believed that "global oil supplies will only meet demand until global oil production has peaked sometime between 2013 and 2020." Afterward, oil production will decrease and create a gap in the global energy market, which can be bridged by unconventional oil and renewable energy sources. On the basis of a compound growth rate, Asif and Muneer (2007) have estimated the years for exhausting coal in India, Russia, and USA as 190, 112, and 84 respectively. These numbers based on a nil growth rate will be 315, 1,034, and 305 years. Based on Shafiee and Topal's (2009) calculation, the depletion time for oil, gas, and coal is estimated at 35, 37, and 107 years. They emphasize that "coal reserves are available up to 2012, and will be the only fossil fuel remaining after 2042." These estimations prove that coal reserves are much larger than oil and gas; therefore, coal will be an important source of energy in the future.

2.2.2 Renewable Energy

Although renewable energy (RE) has been used as a major source of energy for centuries, currently it constitutes only a small percentage of the world's total primary energy supply. According to BP, the share of renewable energy in the global primary energy consumption was 1.6 % in 2011. The USA, Germany, and China have been the biggest consumers of renewable energy sources at 45.3, 23.2, and 17.7 Mtoe respectively. Renewable energy accounted for about half of the estimated 208 GW of new electricity capacity installed in 2011. By region, the EU has the largest nonhydropower capacity, which is 174 GW. The estimated share of renewable energy in global electricity production has been around 20 % (including hydropower). Renewable energy is also used in the form of biofuels in the transportation sector. Liquid biofuels constituted around 3 % of global road transport fuels in 2011 (Martinot and Sawin 2012).

Many countries have started to install facilities in order to use renewable energy sources for power generation. But the share of renewable energy supply varies by region and country. Europe is considered as a front-runner in renewable energy technologies, with RE industry in Europe already reaching an annual turnover of 10 billion euros and employing 200,000 people (Kaygusuz et al. 2007). According to *Renewables 2012 Global Status Report* (Martinot and Sawin 2012), "Significant technology and cost reductions of renewable energy technology, along with improved business and financing models, are increasingly creating clean and affordable renewable energy solution for individuals and communities in developing countries." China, the USA, Brazil, Canada, and Germany were the top five countries in 2011 in terms of their capacity to produce renewable energy electricity. If we consider nonhydroelectric renewable energy power capacity, this ranking is changed to China, the USA, Germany, Spain, and Italy. The fifth ranking in both cases is followed closely by India. China installed 70 GW (mostly wind power) last year, and the country's 282 GW of hydropower generation capacity is not included (Martinot and Sawin 2012).

According to an IEA report (IEA 2012c), renewable energy subsidies sharply increased to 88 billion dollars in 2011, which shows a growth rate of 24 % over 2010. "Government policies have been essential to recent growth in renewable energy, especially in the power sector. Environmental concerns have been a key policy driver, targeting emissions reduction of carbon dioxide and local pollutants. Renewables have also been supported to stimulate economies, enhance energy security and diversify energy supply." It has been mentioned in *GSR 2012* (Martinot and Sawin 2012) that worldwide new investment in renewable energy sources increased to 257 billion dollars in 2011, which is twice the investment in 2007 and six times higher than 2004. Wind and solar energy are the main sources of renewable energy used by many countries. Table 2.3 shows the figures for wind and solar energy consumption over 2010–2011.

As we see in Table 2.3, Europe is the front-runner in using renewable energy technologies; it has the biggest capacity for wind energy and the highest growth rate for solar energy. Also, Europe accounts for almost 42 % of global wind energy consumption. Wind energy usage for Germany, Spain, and the UK in 2011 was 46.5, 42.4, and 15.8 TWh respectively. According to Kaygusuz et al. (2007), "Impressive annual growth rates of more than 40 % between 1996 and 2003 have made Europe into the frontrunner in wind energy technology development." In the Asia Pacific

Table 2.3 Wind and solar energy consumption over 2010–2011 (TWh)

	Wind			Solar		
	2010	2011	Change (%)	2010	2011	Change (%)
North America	105.2	133.3	26.7	1.3	2.1	55.2
S. & Cent. America	3.6	4.4	22.7	a	a	b
Europe & Eurasia	152.5	182.0	19.4	23.2	44.6	92.2
Middle East	0.3	0.3	0.1	a	0.1	99.5
Africa	2.3	2.3	0.8	a	0.1	43.6
Asia Pacific	83.9	115.1	37.2	5.3	8.9	68.2
World	347.8	437.4	25.8	29.9	55.7	86.3

Source: *BP Statistical Review of World Energy, 2012*
Notes: *a* less than 0.05, *b* less than 0.05 %

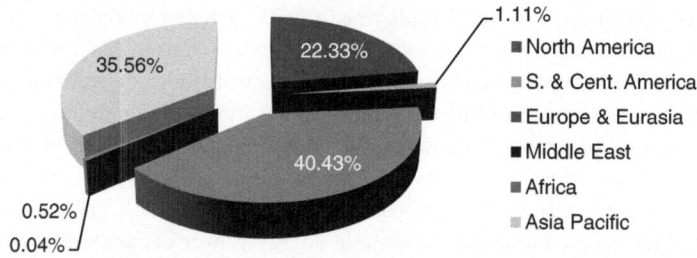

Fig. 2.3 Cumulative installed wind turbine capacity, 2011 (Reproduced from BP (2012))

region, China's consumption in 2011 was 73.2 TWh, constituting more than 60 % of the Asia Pacific market and 16.7 % of the global market.

As an individual country, the USA has the first rank for wind energy consumption (121 TWh), which is comparable with total consumption in the top three consumer countries in Europe (135.1 TWh) and accounts for 27.7 % of the global market. Some development in wind energy has taken place in regions like the Middle East and Africa. Along with Ethiopia, which has joined the ranks of countries that have commercial-scale projects for using wind energy, the South African market is venturing into wind energy. In the Middle East, Iran is the only country with large-scale wind projects, and it had a total of 91 MW at the end of 2011 (Martinot and Sawin 2012). There was little development in Iran over 2010–2011 compared with the previous years' trend due, at least in part, to the imposition of sanctions on Iran and the economic difficulties it faced in developing these projects.

Regarding installed wind turbine capacity, the most significant growth was seen in Argentina and Brazil with 239.4 and 53.8 % respectively. The region of South and Central America has the highest growth rate (66 %) for cumulative installed wind capacity in the world. Almost 67 % of the current global capacity is installed in four countries. China leads the list followed by the USA, Germany, and Spain (BP 2012). The distribution of total installed global wind turbine capacity is shown in Fig. 2.3.

Although Europe is considered as a leader in terms of cumulative installed capacity, there are many installations outside Europe. If we will not experience any technological advance during this time period, then we can forecast that wind power will be able to generate 10–20 % of the global electricity by the year 2050—this has already been achieved in Denmark (Tester et al. 2005).

Solar energy is the second main source to deploy renewable energy. The use of photovoltaic energy is growing quickly. The size of global installed capacity was 2 GW in 2002 compared with 69 GW in 2011 when the solar photovoltaic (PV) had an extremely high growth rate, as was the case a year before. About 30 GW of new capacity has been installed globally, increasing worldwide cumulative installed photovoltaic power by 73 % to 69 GW, and it is almost 10 times the global capacity in 2006 (BP 2012) As we see in Table 2.3, Europe is the major area for using solar energy. Germany is the leader in this region followed by Italy and Spain. Most of the new photovoltaic systems have been installed in Europe, which has almost 74 % of the total capacity in the world. According to BP, the installed capacity in Germany and Italy was 24.8 and 12.8 GW respectively, which constitutes 54 % of the global installed photovoltaic power in 2011. Other top markets in Europe include France, the Czech Republic, Belgium, and the UK. The top five countries for cumulative installed solar PV at the end of 2011 were Germany, Italy, Japan, China, and the USA, closely followed by Spain.

According to the *Global Status Report 2012* (Martinot and Sawin 2012), "For the first time ever, solar PV accounted for more additional capacity than any other type of electricity generating technology: PV alone represented almost 47 % of all new EU electric capacity that came on line in 2011." Although installation of PV power plants shows an extreme growth rate around the world, the size of solar energy consumption in the Middle East and Africa is much lower than other regions. There is a huge potential in these areas for deploying solar energy, but they have not used this source of energy as much as other countries so far as they have rich sources of fossil fuels. Fossil fuels are subsidized in petroleum-exporting countries, accounting for 34 % of the worldwide subsidies. Iran's subsidies at a rate of 82 billion dollars were the highest in 2011 despite the introduction of energy price reforms in 2010. Saudi Arabia has the second-highest subsidies at 61 billion dollars (IEA 2012c). These subsidies are the main reason why these countries fall behind others in the deployment of solar energy. Breyer et al. (2010) have argued that PV power plants have achieved parity with oil power plants on a total cost basis and it is possible for the Middle East and North Africa (MENA) region to reach fuel parity for PV and fossil fuel power plants in the first half of the 2010s.

2.2.3 Outlook of Energy Consumption

It is expected that the global population will increase to 8.6 billion by 2035 (IEA 2012a). Consequently, there will be a growth in economic activities and energy consumption. Of course, some events cannot be forecast in the long term. The Asian financial crisis in 1997–1998 and the US subprime mortgage crisis in 2008–2009 are two examples of shocks for the global economy. Most projections are usually calculated on the basis of gradual trends. Economic growth, energy consumption, and environmental issues are affected by economic shocks. Economic recovery varies among different countries. The recession in the USA has finished officially, but Europe's recovery has a time lag. According to the *International Energy Outlook 2011* reference scenario, most countries will have resumed the economic growth rate forecast for the long term before the crisis by 2015 (Conti and Holtberg 2011). It states that global GDP will increase annually by 3.4 % on average over 2008–2035. This rate is estimated to be 4.6 and 2.1 % for non-OECD and OECD countries respectively. Figure 2.4 shows world energy consumption outlook by groups of countries and the world.

According to *IEO 2011*, world energy consumption will increase by 53 % over the years 2008–2035. The average annual percentage change is 1.6 % globally and is forecast as 0.6 and 2.3 % for OECD and non-OECD countries respectively. Energy consumption in non-OECD countries (led by China and India) shows a phenomenal growth rate of 117 % during the outlook period. China and India will account for 31 % of global energy consumption in 2035. The slowest growth rate among non-OECD countries will be Europe and Eurasia—just 16 % between 2008 and 2035—due to its population decline and energy efficiency achieved by replacing inefficient equipment (Conti and Holtberg 2011).

We should mention that different scenarios in *IEO 2011* are calculated on the basis of oil prices and energy demand. For example, alternative energy supply conditions are forecast on the basis of high and low oil prices. Also, the impact of high and low non-OECD demand on the global market is estimated. *World*

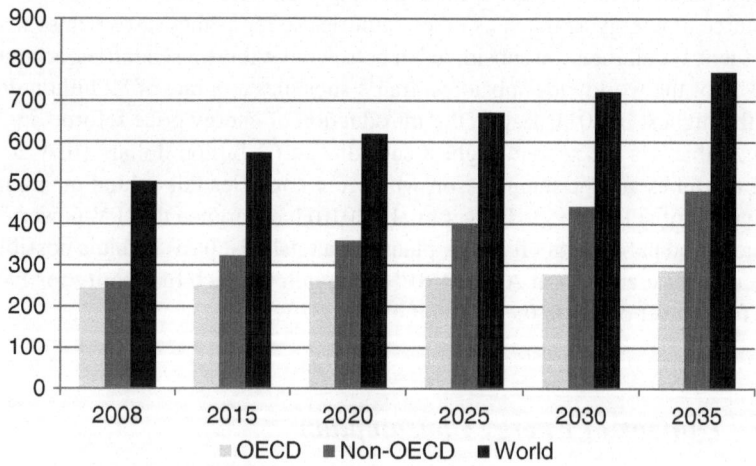

Fig. 2.4 World energy consumption by country group (quadrillion British thermal unit, QBtu) (Reproduced from Conti and Holtberg (2011))

Energy Outlook (*WEO*) scenarios are defined by the underlining assumption about government policies. In this regard, four scenarios are differentiated by the IEA in the *WEO 2012* report: Current Policies Scenario, New Policies Scenario, 450 Scenarios, and Efficient World Scenario. Nonpolicy assumptions are economic growth, population, and energy prices, which are considered for each scenario.

New Policies Scenario, which is called the central scenario or reference in the IEA report, considers all policies and commitments already implemented alongside those policies that have been announced and are to be introduced. Current Policies Scenario includes those government policies that had been made as a law or implemented by mid-2012 without considering any possible policy in the future. 450 Scenarios is defined on the basis of the possibility of limiting the increase in global average temperature to 2 °C compared with preindustrial levels. Experts believed that GHGs should be limited to 450 ppm of carbon dioxide equivalent in order to meet this target. Efficient World Scenario quantifies the implication of major changes in energy efficiency for the economy, the environment, and energy security (IEA 2012c). Figure 2.5 shows the total primary energy demand (TPED) based on country grouping in the New Policies Scenario.

According to the IEA estimation, OECD energy demand in 2035 will be 3 % more than 2010, but fuel substitution will make some changes in the energy mix. The OECD oil and coal demand is forecasted to decrease over 2010–2035 by 21 and 24 % respectively. By contrast, the share of natural gas and renewable energy is rising. The biggest change is related to renewable energy, which will make for around 33 % of OECD power generation in 2035. Although the share of nuclear power has increased from 19 % to 21 % due to some changes made by Europe and Japan to reduce their reliance on nuclear power, it will increase in absolute figures due to nuclear generation growth in North America and South Korea (IEA 2012c).

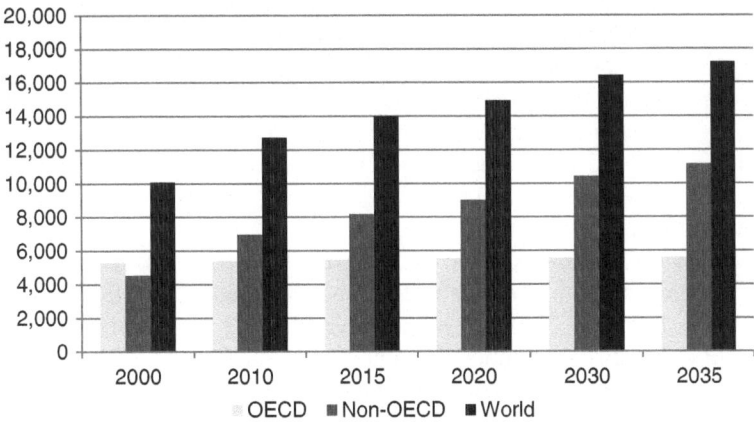

Fig. 2.5 Total primary energy demand by country group (Mtoe (Reproduced from IEA (2012c))

2.3   Energy Consumption and Economic Growth

The level of economic activities plays a key role in energy consumption and is considered a key driver of energy markets. Many scholars have studied the relation between energy consumption and economic growth. Generally, four types of hypotheses are defined in this regard. If there is no causality, called neutral hypothesis, it means energy consumption is not related to GDP. Therefore, neither conservative nor expansive policies affect economic growth. Unidirectional causality may exist from economic growth to energy consumption (conservation hypothesis) or energy consumption to economic growth (growth hypothesis). Feedback hypothesis is applicable when there is bidirectional causality. Depending on each hypothesis, energy policies have different influences on economic growth. Discussion about the impact of energy consumption on economic activities got importance after the Arab oil embargo in 1973.

Early research in this regard was published in the 1970s. Allen et al. (1976) projected economic growth and energy demand for the USA over 1975–2010. Hitch (1978) discussed how much energy consumption conservation can contribute to energy supply and how it influences economic growth. The idea of causality relationship between energy consumption and economic growth was introduced by Kraft and Kraft (1978). They used the Granger causality test to define the relation between gross energy inputs and gross national product (GNP) and found that causality is unidirectional, running from GNP to energy for the postwar period in the USA. Akarca and Long (1980) and Yu and Hwang (1984) applied Sim's method by using US data and found no causal relationship between GNP and energy. Furthermore, Yu and Hwang argued that there is a slight unidirectional relation from employment to energy consumption. Yu and Jin (1992) showed that a long-run equilibrium relationship does not exist between energy consumption and real output or employment in the USA. Stern (1993) examined the causal relationship between GDP and energy use applying the vector autoregressive (VAR) model. He argued that the results of the Granger test are different for measuring impacts of quality-weighted final energy use and gross energy use on GDP. The former shows a causal relationship running from energy consumption to economic growth, but in the latter, it is vice versa. According to his research, conservative energy policy and rising tax on energy without specifying the ways for energy saving may reduce economic growth.

Cheng (1995) reexamined the causality between energy consumption and economic growth with both bivariate and multivariate models for US data over the period 1947–1990. According to his research, there is no causality relationship from GNP to energy consumption. In another research, Cheng (1998) used Hsiao's Granger causality and found that employment and real GNP directly cause energy consumption. Based on his findings, energy conservation policy may not affect a country like Japan. Also, Cheng (1999) applied the Johansen cointegration test to investigate this relationship for India and detected no causality from energy consumption to economic growth. He found that causality runs from economic growth to energy consumption instead. Stern (2000) extended his previous work on the analysis of the causal relationship between GDP and energy use in the USA for the postwar period by using the cointegration test, and his findings were similar to the Granger causality results.

In recent works, Wolde-Rufael (2005) examined the long-run relationship between energy use per capita and GDP per capita for 19 African countries applying two methodologies—the developed cointegration test proposed by Pesaran, Shin, and Smith and the Toda and Yamamoto test. His research showed a long-run relationship between energy use and GDP per capita for 8 of the 19 countries and a causal relationship for 12 countries. Lee and Chang (2008) applied panel models to reinvestigate the comovement causal relationship within a multivariate framework for 16 Asian countries. Their results indicate that energy consumption is caused by GDP in the long term but not vice versa, and there is no short-term or long-term relationship from GDP to energy consumption. It means more energy consumption comes with higher GDP, but it is not the same from GDP to energy consumption. Lee and Chang (2008) further divided the 16 countries into Asia-Pacific Economic Cooperation (APEC) and Association of Southeast Asian Nations (ASEAN) members. Their findings strongly support that energy consumption has a significant impact on economic growth in Asian countries. Therefore, continuous energy use can generate a continuous increase in economic output. In other words, GDP is fundamentally driven by energy.

Narayan and Prasad (2008) used a bootstrapped approach to causality for testing the mutual impact of electricity consumption and GDP for 30 OECD countries. They found causality from electricity consumption to GDP for eight countries. This means the electricity conservation policy has a negative effect on real GDP in these countries. But this policy does not influence the other 22 countries. Narayan and Prasad (2008) also indicated that real GDP causes electricity consumption for six countries and policymakers should have strategies to ensure enough energy supply to achieve the planned economic growth rate. Chontanawat et al. (2008) examined the causal relationship from energy consumption to GDP for 30 OECD and 78 non-OECD countries. They found that causality from energy to GDP in OECD countries is more prevalent than non-OECD countries, implying that energy conservative policies have a greater impact on the economic growth of developed countries than developing countries.

Huang et al. (2008) used a panel data of energy consumption and GDP for 82 countries to investigate causality. They classified these countries into four groups based on income levels defined by the World Bank: low-income group, lower middle-income group, upper middle-income group, and high-income group. According to their findings, using data for all countries as one group shows a bidirectional positive relationship between economic growth and energy consumption. But the result is different when the method is applied for different groups. Huang et al. (2008) detected a unidirectional positive relationship from economic growth to energy consumption for the middle-income group countries and a negative one for the high-income group countries.

Apergis and Payne (2010a) investigated the causal relationship between renewable energy consumption and economic growth for a panel of 20 OECD countries, applying the panel cointegration and error correction model. According to their findings, the short-run and long-run Granger tests detected positive bidirectional causality between renewable energy consumption and economic growth. Also, renewable energy influences economic growth because of its positive effect on the real gross fixed capital formation. Apergis and Payne (2010a) argued that this evidence proves the importance of renewable energy sources in the energy portfolio of the OECD countries. The estimation of the vector error correction model shows both short-run and long-run bidirectional causality between renewable energy consumption and economic growth. They indicated that this result emphasizes the benefits associated with supportive policies for renewable energy such as tax credits on production, rebate for the system installation, portfolio standards, and markets for renewable energy certificates.

In another research, Apergis and Payne (2010b) examined the causal relationship between real GDP, renewable energy consumption, real gross fixed capital formation, and labor force for 13 countries within Eurasia. Due to the importance of Russia in the Eurasia region, they categorized two data sets to run the causality test with and without it. The result of panel cointegration tests for both data sets shows a long-run relationship between real GDP, renewable energy consumption, real gross fixed capital formation, and labor force. The result of panel error correction models shows both a short-run and a long-run bidirectional causal relationship between renewable energy consumption and economic growth. Apergis and Payne (2010b) indicated that a multilateral effort to develop renewable energy and energy efficiency should be encouraged by policymakers. Also, they stated that a proper incentive mechanism to promote market availability of renewable energy should be introduced.

Wolde-Rufael and Menyah (2010) tried to test the causal relationship between nuclear energy consumption and real GDP for nine advanced countries applying the Toda and Yamamoto version of the Granger causality test. They found a unidirectional causality running from nuclear energy consumption to economic growth in Japan, the Netherlands, and Switzerland; a unidirectional causality from economic growth to nuclear energy consumption in Canada and Sweden; and a bidirectional causality in France, Spain, the UK, and the USA. Since the causality relationship in France, Japan, the Netherlands, and Switzerland is negative, they argued, energy conservative policies could help mitigate the negative effects of increasing nuclear energy consumption on economic growth. Lee and Chiu (2011) applied four methodologies—the Johansen cointegration test, the Granger noncausality test, the generalized impulse response function, and the generalized forecast error variance decomposition—to investigate the relationship between nuclear energy consumption, real oil price, oil consumption, and real income in six highly industrialized countries. Their results show a unidirectional causality running from economic growth to nuclear energy consumption in Japan. It means the conservation energy policy does not influence economic growth. Also, there is a bidirectional relationship between nuclear energy consumption and real income in Canada, Germany, and the UK, but no causality was found between these two parameters in France and the USA.

Unlike previous studies, Apergis and Payne (2012) investigated the simultaneous consumption of renewable and nonrenewable energy in order to examine the causal relationship between them and economic growth for 80 countries. According to their findings, there is a bidirectional causality between renewable and nonrenewable energy consumption and economic growth in both short-run and long-run periods. It means both types of energy sources are important for economic growth. Furthermore, the results show a negative bidirectional causality between these measures, implying substitutability of renewable and nonrenewable energy sources. Apergis and Payne (2012) argued that substitutability of renewable and nonrenewable energy sources supports continuation of governmental policies to promote renewable energy consumption as well as implementation of policies to reduce nonrenewable energy consumption.

Yildirim and Aslan (2012) applied both the Toda–Yamamoto procedure and the bootstrap-corrected causality test for 17 highly developed OECD countries to investigate the relationship between energy consumption, economic growth, employment, and gross fixed capital formation. They found a bidirectional causality between energy consumption and real GDP for Italy, New Zealand, Norway, and Spain. The authors believed that due to the support feedback hypothesis for these countries, the energy conservation policy should not be followed by these countries at the aggregated level because the total economy is influenced by opposite effects. It means economic growth will be reduced by lower levels of energy consumption and vice versa, making for a circular relationship. In this situation, the energy policy should be regulated carefully and diversified on the basis of sectors or energy kinds. According to the findings of this study, there is a unidirectional causal relationship from energy consumption to economic growth for Japan and in the opposite direction for Australia, Canada, and Ireland, whereas there is no causality relationship for all the other nine countries. Yildirim and Aslan (2012) also tested the importance of lag length in their research and found that the selection of lag length is important for Denmark, Ireland, Norway, and Spain. Table 2.4 compares the results of these empirical studies.

The causality relationship between energy consumption and economic growth is analyzed in order to examine the possible effects made by energy policies. As is evident from Table 2.4, there may be different results for some countries in the same period with different methodologies or even similar methodologies. Also, it should be taken into account that this analysis considers individual relationships between two variables (in this case, energy consumption and economic growth). Therefore, such analysis of effects is not reliable to be a basis for making decisions regarding energy policy. There are other parameters such as technological innovation and governmental taxes that may affect this relationship. The impacts of energy policies are conditional on the country, applied methodology, and time effects in the sample of data. Also, the interaction effects with other variables should be considered.

Table 2.4 Comparing empirical studies on the energy consumption–growth nexus

Author	Period	Country	Methodology	Causality relationship
Kraft and Kraft (1978)	1947–1974	USA	Granger causality	GNP → EC
Akarca and Long (1980)	1950–1970	USA	Sim's technique	Neutral
Yu and Hwang (1984)	1947–1979	USA	Sim's technique	Neutral
Yu and Jin (1992)	1974–1990	USA	Co-integration, Granger	Neutral
Stern (1993)	1947–1990	USA	Multivariate VAR model	EC → GDP
Cheng (1995)	1947–1990	USA	Co-integration, Granger	Neutral
Cheng (1998)	1952–1995	Japan	Hsiao's Granger causality	GNP → EC
Cheng (1999)	1952–1995	India	Co-integration, ECM, Granger	GDP → EC
Stern (2000)	1948–1994	USA	Co-integration, Granger	EC → GDP
Wolde-Rufael (2005)	1971–2001	19 African countries	Co-integration, modified Granger	GNP → EC (5)
				EC → GNP (3)
				GNP ←→ EC (2)
				Neutral (9)
Lee and Chang (2008)	1971–2002	16 Asian countries	Co-integration, ECM	EC → GDP
Narayan and Prasad (2008)	1960–2002	30 OECD	Bootstrapped causality	EC → GDP (8)
				GDP → EC (22)
Chontanawat et al. (2008)	1960–2000	30 OECD 78 Non-OECD	Co-integration, Granger	EC → GDP (21 OECD, 36 non-OCED)
Apergis and Payne (2010a)	1985–2002	20 OECD	Co-integration, ECM	GDP ←→ RE
Apergis and Payne (2010b)	1992–2007	13 Eurasia	Co-integration, ECM	GDP ←→ RE
Wolde-Rufael (2005)	1971–2005	9 developed countries	Modified Granger	NE → GDP (3)
				GDP → NE (2)
				GDP ←→ NE(4)
Lee and Chiu (2011)	1965–2008	6 highly industrialized countries	Co-integration, Granger, generalized impulse response function	GDP → NE(1)
				Neutral (2)
				GDP ←→ NE(3)

(continued)

Table 2.4 (continued)

Author	Period	Country	Methodology	Causality relationship
Apergis and Payne (2012)	1990–2007	80 countries	Co-integration, ECM	RE, NRE ←→ GDP
Yildirim and Aslan (2012)	1971–2009	17 highly developed OECD	Bootstrap-corrected test, modified Granger	EC → GDP (1)
				GDP → EC (3)
				GDP ←→ EC (4)
				Neutral (9)

Notes: *VAR* vector autoregressive model, *ECM* error correction model, *EC* energy consumption, *GDP* gross domestic product, *RE* renewable energy consumption, *NRE* non-renewable energy consumption, *NE* nuclear energy consumption

EC → GDP means that the causality runs from energy consumption to economic growth

GDP → EC means that the causality runs from economic growth to energy consumption

EC ←→ GDP means there is a bi-directional causality between energy consumption and economic growth

Neutral means there is no causality between energy consumption and economic growth

2.4 The Main Drivers for Using Renewable Energy

The first driver for seeking alternative energy sources has been energy security since the Arab oil embargo in 1973 or the first oil shock. The oil shocks in the 1970s stimulated interest in renewable energy sources. The global concern about climate change and sustainability encouraged countries to invest in renewable energies. We can define three main drivers for using renewable energy: energy security, economic impacts, and CO_2 emission reductions.

2.4.1 Energy Security

As we have mentioned, concerns about the security of energy supply rose after the Arab oil embargo in 1973. There are other factors such as high oil prices, increasing dependency on oil imports, depletion of fossil fuels, increasing competition from emerging economies, political instability in major oil-producing countries, and the high impact of any disruption in energy supply on developed and rapidly developing countries (Bhattacharyya 2011). The level of insecurity is reflected by the risk of supply disruption and the estimated cost incurred for making security. Owen (2004) called security of energy supplies a key requirement for the economic, environmental, and social objectives of sustainable development policies. In his view, energy security risk could be classified into strategic and domestic system risks. He also defined damage costs and control costs as potential costs imposed by energy insecurity. He argued that damage costs could be evaluated by a potential decrease in GNP but it is difficult to estimate how much money is spent as control costs on energy security. For example, it is very difficult to estimate how much money has been spent by the USA to control oil security.

Delucchi and Murphy (2008) investigated the impact of US military costs on motor vehicle fuels and estimated that in the case of no oil in the Persian Gulf, US defense expenditure might be reduced by about $27–$37 billion per year, meaning $0.03–$0.15 per gallon ($0.01–$0.04 per liter). Hedenus et al. (2010) analyzed the expected economic cost of oil supply disruption by energy policies in the EU-25. They analyzed how energy policies affect the oil market and how much money could be gained by these policies. The results show that the expected cost of oil disruption is 29.5–31.6 billion euros a year, corresponding to roughly 9–22 euros/bbl or 6–14 c/l of gasoline. Delucchi and Murphy (2008) also estimated GHG benefits of 20 euros/ton carbon dioxide to substitute oil for pellets in the residential sector for heating.

Concerns about climate change have made energy security objectives more crucial. The diversification of energy supply to promote energy security could be considered as a policy for climate protection (Bhattacharyya 2011). Before the industrialization era and prior to using coal as a main source of energy in the mid-nineteenth century, renewable energy sources were used widely. There are huge potential sources of renewable energy such as hydropower, solar, wind, and biomass around the world, which are able to supply clean energy and enhance long-term sustainable energy supply (Asif and Muneer 2007). Renewable energy sources may have security issues due to intermittent characteristics of some kinds of energy such as solar and wind energy or the possibility of low rainfall for hydropower consumption. Therefore, such factors should be considered for the sectors that rely heavily on these sources. Renewable energy technologies are beneficial for countries that produce and consume energy. Renewable energy technologies reduce domestic demand for fossil fuels and increase the capability for export. For example, Iran was the fourth largest worldwide natural gas producer in 2011 but was a net importer because of high domestic demand. Also, high dependency on imports could pose serious problems if there is any disruption in energy supply. For example, European countries are dependent on Russia for importing natural gas. They experienced a difficult situation when Russia cut off all gas supply transmitted by Ukraine in 2006.

Generally, renewable energy technologies are considered as an expensive option, which are not compatible with traditional sources of energy, but some technologies like wind power are more feasible today, and the cost of other technologies such as solar photovoltaic is decreasing rapidly (IEA 2011). Furthermore, we should consider external costs that are spent for energy security indirectly in our calculation. Alongside storage costs and military expenditure, there is an extreme externality cost due to the possibility of accidents in nuclear power plants such as Three Mile Island (1979), Chernobyl (1986), and Fukushima Daiichi (2011). According to a report prepared for the international organization Chernobyl Forum (2003–2005), the total amount spent by Belarus over 1991–2003 is evaluated at more than 13 billion dollars. Also, the total losses over 30 years have been estimated at around 235 billion dollars by Belarus (Chernobyl Forum 2006). According to this report, in Ukraine 5–7 % of government expenditure is still allocated to Chernobyl-related programs. Around 6,000 thyroid cancers have been found in contaminated regions of the Chernobyl accident to date, and an additional 10,000–40,000 cases of cancer are estimated over the next decades (Ten Hoeve and Jacobson 2012). The number of accidents in nuclear power plants may be rare, but there will be an extreme cost in terms of economic, social, and environmental aspects. If we include all external costs including social and environmental security in our evaluation, renewable energy sources will be feasible.

2.4.2 Economic Impacts

The emphases for economic impacts are job creation, industrial innovation, and balance of payment. Renewable energy technologies could enable countries with good solar or wind resources to deploy these energy sources to meet their domestic demand. In parallel, demand management policies are used in energy areas to reduce the demand through various energy-saving technologies and policies (Heshmati 2014). Also, renewable energy technologies may even enable these countries to deploy renewable energy sources with long-term export potential. Also, the cost of importing fuels can affect economic growth. Some major consumer countries like the USA have domestic resources that enable them to cover a part of the demand. The USA spent around 410 billion dollars in 2008 to import fossil fuels, constituting more than 3 % of its GDP, but this figure could be higher for developing countries without enough energy resources (IEA 2011) Therefore, if these countries could reduce their balance of payment by producing renewable energy to replace fossil fuels, they could make a capacity for investment in other sections. The IEA made a cost–benefit analysis for investment in low-carbon energy systems based on two scenarios: ETP 2012 6 °C (6DS), which assumes business as usual, and 2 °C (2DS), which is targeted at reducing carbon dioxide emissions by 50 % compared to 2005 levels. The results show that an estimated 103 trillion dollars will be saved during the years 2010–2050 by reducing fossil fuel consumption. It is indicated that this calculation is based on reduction in fossil fuel purchase (214 Gtoe) and it could be 150 trillion dollars if the impact of lower fuel prices is taken into consideration (IEA 2012a).

A main economic driver to enhance renewable energy technologies is their potential to create jobs. It is estimated that 5 million people work in renewable energy industries. Although total employment in these industries continues to increase, some countries such as Germany and Spain suffered recently because of the global recession and policy changes (Martinot and Sawin 2012). Figure 2.6 shows the distribution of estimated jobs in renewable energy worldwide by industry based on the *GSR 2012* report.

The *GSR 2012* has estimated the breakup of job creation by sector as follows: 1.5 million workers in biofuels, 820,000 in solar PV, and 670,000 in wind power. Currently, more than 1.6 million workers are employed in the renewable energy industry (Martinot and Sawin 2012). The majority of jobs in renewable energy

Fig. 2.6 Estimated job in renewable energy worldwide, by industry (Reproduced from Martinot and Sawin (2012))

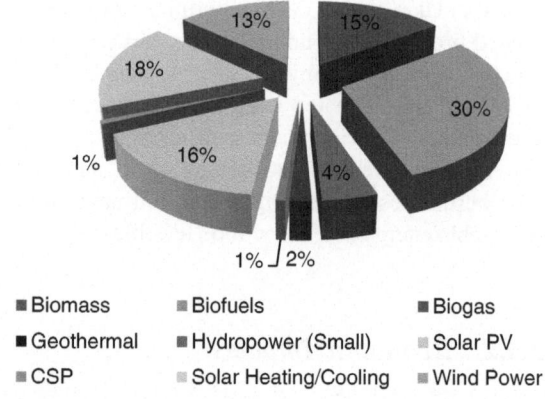

- ■ Biomass   ■ Biofuels   ■ Biogas
- ■ Geothermal   ■ Hydropower (Small)   ■ Solar PV
- ■ CSP   ■ Solar Heating/Cooling   ■ Wind Power

industries are located in China, Brazil, the USA, and the EU. Germany is the front-runner country in Europe for job creation in the renewable energy industry. Germany has increased power generation sharply by renewable technologies since the beginning of this century with a share of almost 15 % of total electricity production in 2008 (Frondel et al. 2010). Ragwitz et al. (2009) investigated the gross and net effects of renewable energy policies in the EU and analyzed the past, present, and future effects of renewable energy policies on employment and economics of countries in general and at the members' levels. They found that current high economic benefits of renewable energy sectors could be increased in the future "if the current policies are improved in order to reach the agreed target of 20 % renewable energies in Europe by 2020." They argued that increasing share of renewable energy sources not only has no negative effect on the economy but could also help the economy by job creation and increasing GDP. In their view, the economic advantage of renewable energy could be higher if external costs are included.

Mathiesen et al. (2011) examined a 100 % renewable energy system including transport by the year 2050 and considered two short-term transition targets in 2015 and 2030. They also indicated that implementing renewable energy technologies could have positive socioeconomic impacts including job creation and increasing exports. Several market leaders including Germany, Denmark, and Japan have focused on industrial and economic development objectives to support renewable energy technologies through stable policy frameworks, innovation chains, and a good environment for investment. They have specialized in the knowledge-based stage and became front-runners in terms of innovation in the renewable energy industry. This situation gives them a first-mover advantage in global renewable energy trade and technology development (IEA 2011). "International trade performance depends on technological capability. If a country has a comparatively high knowledge base, it also has an additional advantage in developing and marketing future technologies" (Walz et al. 2009).

2.4.3 CO₂ Emission Reduction

Renewable energy technologies could reduce carbon dioxide emissions by replacing fossil fuels in the power generation industry and transportation sector. Life-cycle CO_2 emissions for renewable energy technologies are much lower than fossil fuels. The life-cycle balance is also considered an important factor for heat and transportation sectors. Based on an analysis performed by IEA, renewable power generation enabled focused countries to save 1.7 Gt CO_2 emissions in 2008, which is more than the total power sector's CO_2 of the Europe region (1.4 Gt) (Ölz 2011). This analysis shows that hydropower technology constitutes the largest share for saving CO_2 emissions with 82 % followed by biomass and wind with 8 and 7 % respectively.

According to an IEA analysis, the potential saving of OECD and BRICS countries (Brazil, Russia, India, China, and South Africa) is estimated roughly at 5.3 Gt in 2030, almost the same as forecast for power-related CO_2 emissions in *WEO 2010* for these countries in 2030 under a 450 ppm scenario (5.8 Gt).

Figure 2.7 shows the CO_2 saving under the WEO 450 scenario compared with the no-RE scenario in 2030. The key point is that the largest CO_2 saving is concentrated in OECD countries and China. According to an IEA report, CO_2 saving in China on a 450 ppm scenario would be 2.2 Gt, constituting 64 % of the BRICS total saving (Ölz 2011). Edenhofer et al. (2010) examined technological feasibility and economic consequences of achieving GHG targets and found that these targets are low enough to be feasible technically and economically. They stated that this viability crucially depends on particular technologies. For example, the availability of carbon capture storage technology is very important to remove CO_2 from the atmosphere. They also argued that additional political and institutional prerequisites are required to achieve the targets.

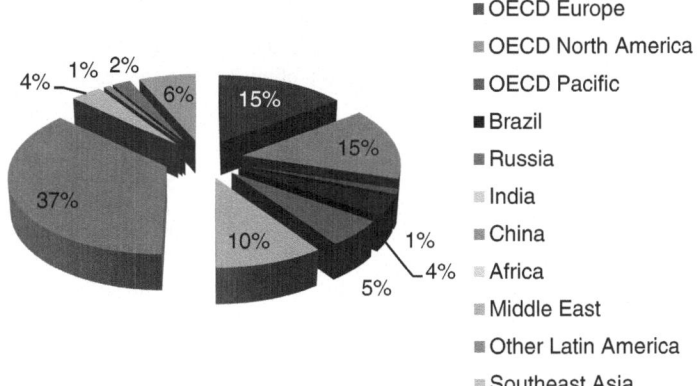

Fig. 2.7 Saving in CO_2 emissions between no-RE and 450 scenarios in 2030 (Reproduced from Ölz (2011))

References

Akarca AT, Long TV (1980) Relationship between energy and GNP: a reexamination. J Energy
    Develop 5(2):326–331
Allen EL, Cooper CL, Edmonds FC, Edmonds JA, Reister DB, Weinberg AM, Whittle CE, Zelby
    LW (1976) US energy and economic growth, 1975–2010, Institute for Energy Analysis, Oak
    Ridge
Apergis N, Payne JE (2010a) Renewable energy consumption and economic growth: evidence from
    a panel of OECD countries. Energy Policy 38(1):656–660
Apergis N, Payne JE (2010b) Renewable energy consumption and growth in Eurasia. Energy
    Economics 32(6):1392–1397
Apergis N, Payne JE (2012) Renewable and non-renewable energy consumption-growth nexus:
    evidence from a panel error correction model. Energy Economics 34(3):733–738
Asif M, Muneer T (2007) Energy supply, its demand and security issues for developed and
    emerging economies. Renew Sust Energ Rev 11(7):1388–1413
Bhattacharyya SC (2011) Energy economics: concepts, issues, markets and governance. Springer,
    London
BP (2012) BP Statistical Review of World Energy. www.bp.com/statisticalreview/
Breyer C, Gerlach A, Beckel O, Schmid J (2010) Value of solar PV electricity in MENA
    region. Paper presented at the Energy Conference and Exhibition (EnergyCon), 2010 IEEE
    International
Cheng BS (1995) An investigation of cointegration and causality between energy consumption and
    economic growth. J Energy Dev 21(1):73–84
Cheng BS (1998) Energy consumption, employment and causality in Japan: a multivariate
    approach. Indian Econ Rev 33(1):19–29
Cheng BS (1999) Causality between energy consumption and economic growth in India: an
    application of cointegration and error-correction modeling. Indian Econ Rev 34(1):39–49
Chernobyl's Legacy: Health, Environmental and Socio-economic Impacts and Recommendations
    to the Governments of Belarus, the Russian Federation and Ukraine (2006) Chernobyl Froum.
    IAEA, Vienna
Chontanawat J, Hunt LC, Pierse R (2008) Does energy consumption cause economic growth?:
    evidence from a systematic study of over 100 countries. J Policy Model 30(2):209–220
Conti J, Holtberg P (2011) International Energy Outlook 2011. US Energy Information Adminis-
    tration, EIA. http://www.eia.gov/
Delucchi MA, Murphy JJ (2008) US military expenditures to protect the use of Persian Gulf oil
    for motor vehicles. Energy Policy 36(6):2253–2264
Edenhofer O, Knopf B, Barker T, Baumstark L, Bellevrat E, Chateau B, Kypreos S (2010) The
    economics of low stabilization: model comparison of mitigation strategies and costs. Energy J
    31(1):11–48
EIA U.S. What drives crude oil prices? U.S. Energy Information Administration. Retrieved 18 Apr
    2013, from http://www.eia.gov/finance/markets/
Frondel M, Ritter N, Schmidt CM, Vance C (2010) Economic impacts from the promotion of
    renewable energy technologies: the German experience. Energy Policy 38(8):4048–4056
Hedenus F, Azar C, Johansson DJ (2010) Energy security policies in EU-25—the expected cost of
    oil supply disruptions. Energy Policy 38(3):1241–1250
Heshmati A (2014) Demand, customer base-line and demand response in the electricity market: a
    survey. J Econ Surv 28(5):862–888
Hitch CJ (1978) Energy conservation and economic growth, AAAS Selected Symposium 22.
    Published by Westview Press for the American Association for the Advancement of Science,
    Boulder
Huang BN, Hwang M, Yang C (2008) Causal relationship between energy consumption and GDP
    growth revisited: a dynamic panel data approach. Ecol Econ 67(1):41–54
IEA (2011) Deploying renewables. OECD Publishing, Paris

IEA (2012a) Energy technology perspectives 2012. OECD Publishing, Paris

IEA (2012b) CO2 emissions from fuel combustion 2012. OECD Publishing, Paris

IEA (2012c) World energy outlook 2012. OECD Publishing, Paris

Kaygusuz K, Yüksek Ö, Sari A (2007) Renewable energy sources in the European Union: markets and capacity. Energy Sources B Econ Plan Policy 2(1):19–29

Kraft J, Kraft A (1978) Relationship between energy and GNP. J Energy Dev 3(2):401–403

Lee CC, Chang CP (2008) Energy consumption and economic growth in Asian economies: a more comprehensive analysis using panel data. Resour Energy Econ 30(1):50–65. http://dx.doi.org/10.1016/j.reseneeco.2007.03.003

Lee CC, Chiu YB (2011) Nuclear energy consumption, oil prices, and economic growth: evidence from highly industrialized countries. Energy Economics 33(2):236–248

Lund H (2010) Renewable energy systems: the choice and modeling of 100% renewable solutions. Academic, Burlington

Martinot E, Sawin J (2012) Renewables global status report. Renewables 2012 Global Status Report, REN21. http://www.martinot.info/REN21_GSR2012.pdf

Mason JE (2007) World energy analysis: H2 now or later? Energy Policy 35(2):1315–1329

Mathiesen BV, Lund H, Karlsson K (2011) 100% Renewable energy systems, climate mitigation and economic growth. Appl Energy 88(2):488–501

Narayan PK, Prasad A (2008) Electricity consumption–real GDP causality nexus: evidence from a bootstrapped causality test for 30 OECD countries. Energy Policy 36(2):910–918

Ölz S (2011) Renewable energy policy considerations for deploying renewables

Owen AD (2004) Oil supply insecurity: control versus damage costs. Energy Policy 32(16):1879–1882

Ragwitz M, Schade W, Breitschopf B, Walz R, Helfrich N, Rathmann M, Haas R (2009) The impact of renewable energy policy on economic growth and employment in the European Union. European Commission, DG Energy and Transport, Brussels

Salameh MG (2003) Can renewable and unconventional energy sources bridge the global energy gap in the 21st century? Appl Energy 75(1):33–42

Shafiee S, Topal E (2009) When will fossil fuel reserves be diminished? Energy Policy 37(1):181–189

Stern DI (1993) Energy and economic growth in the USA: a multivariate approach. Energy Economics 15(2):137–150

Stern DI (2000) A multivariate cointegration analysis of the role of energy in the US macroeconomy. Energy Economics 22(2):267–283

Ten Hoeve JE, Jacobson MZ (2012) Worldwide health effects of the Fukushima Daiichi nuclear accident. Energy Environ Sci 5(9):8743–8757

Tester JW, Drake EM, Driscoll MJ, Golay MW, Peters WA (2005) Sustainable energy: choosing among options. The MIT Press, Cambridge, MA

Walz R, Helfrich N, Enzmann A (2009) A system dynamics approach for modelling a lead-market-based export potential. Working Paper Sustainability and Innovation No. S 3/2009, Fraunhofer, ISI

Wolde-Rufael Y (2005) Energy demand and economic growth: the African experience. J Policy Model 27(8):891–903

Wolde-Rufael Y, Menyah K (2010) Nuclear energy consumption and economic growth in nine developed countries. Energy Economics 32(3):550–556

Yildirim E, Aslan A (2012) Energy consumption and economic growth nexus for 17 highly developed OECD countries: further evidence based on bootstrap-corrected causality tests. Energy Policy 61:986–993

Yu ES, Hwang BK (1984) The relationship between energy and GNP: further results. Energy Economics 6(3):186–190

Yu ES, Jin JC (1992) Cointegration tests of energy consumption, income, and employment. Resour Energy 14(3):259–266

Chapter 3
Alternative Renewable Energy Production Technologies

3.1 Introduction

The importance of alternative energy sources is underlined by climate change issues triggered through the excessive use of fossil fuels. Heshmati (2014) provides a comprehensive empirical survey of the ramification of a green economy in self-contained form and accessible to specialists and nonspecialist readers. The three drivers for stimulating renewable energy deployment are energy security, economic impacts, and carbon dioxide emission reductions. These drivers should stimulate two changes in a large number of countries (IEA 2012d). First, the output of renewable energy production is expected to reach 2,167 GW in 2017 from 1,454 GW in 2011 (IEA 2012c). Second, the cost of renewable energy production is expected to go down such that it becomes competitive with the production costs of other existing energy production types. According to IEA, wind is the most competitive type of renewable energy source if conditions (e.g., financing, CO_2 emission levels, and fossil fuel prices) are favorable (OECD 2010).

There are two concepts of clean energy technologies: energy supply technologies and energy efficiency technologies. The energy supply technologies are used for generating electricity from renewable energy sources such as wind and solar power. The energy efficiency technologies are used to enhance energy efficiency. Examples are combined heat and power plants (CHP), virtual power plants (VPP), and smart meters for metering the consumption accurately.

It should be noted that transforming the energy sector and replacing conventional energy with renewable energy is an evolutionary process associated with technological change and market formation. Jacobsson and Bergek (2004) indicated that the transforming process for renewable energy such as wind and solar will only happen after 2020 although the growth rate of consumption has strongly increased during this decade. Besides, renewable energy markets are not formed easily due to the cost disadvantage and the subsidization of fossil fuels.

© Springer Science+Business Media Singapore 2015
A. Heshmati et al., *The Development of Renewable Energy Sources and its Significance for the Environment*, DOI 10.1007/978-981-287-462-7_3

3.2   Renewable Energy Supply Technologies

Renewable energy supply is increasing continuously. A large amount of investment has been made during the recent years, and advanced technologies have enabled countries to produce renewable energy more cost effectively. It is forecast that the number of countries producing over 100 MW renewable energy will increase significantly by 2017 (IEA 2012d). Due to some negative and irreversible externalities coming with conventional energy productions, it is necessary to promote and develop renewable energy supply technologies. As we mentioned earlier, these technologies may be incompatible compared with conventional fuels in view of unit production cost. However, they could be compatible if we consider their associated externalities such as environmental and social effects. Besides, it should be noted that economies of scale could play a key role in reducing the unit production cost. Transmission and distribution costs and technologies do not differ much between the two main types of energies.

3.2.1   Hydropower Technologies

Hydropower is the largest renewable energy source for power generation around the world. Hydroelectricity generation has increased strongly over the past 50 years. It was 340 TWh in 1950 and covered about one-third of the global demand. It increased to 1,500 TWh in 1975 and further to 2,994 in 2005. We can compare it with the global consumption of 15,000 TWh and the global production of 18,306 TWh in 2005 (Ngô and Natowitz 2009). Currently, hydropower development is difficult due to its large initial investment (high capital cost despite its low operational cost) and the environmental concerns it raises. It creates some problems for local residents because of relocation of large populations. Also, considering that building dams is permanent and the sunk cost of utilities cannot be removed, the environment is influenced by hydropower construction because of large engineering works. On the other hand, hydropower is attractive because it can supply water for agriculture, household, and industrial use; it is clean; and it has the ability to store energy. Also, it can be used for the application of both base load and peak load power plants.

The largest worldwide capacity hydropower plant is Itaipu installed in Parana River, developed jointly by Brazil and Paraguay. The initial capacity was 12.6 GW in 1984 and extended to 14 GW in 2006 (Ngô and Natowitz 2009). Many argue that hydropower plant construction projects could improve local economies. For example, the USA employed hundreds of workers in the Hoover dam project during the depression in the 1930s (Tester et al. 2005). Hydropower plays a key role for some countries such as Norway. According to the BP statistics (2012), hydroelectricity demand in Norway (122 TWh) constituted almost 64 % of the primary energy consumption in 2011 compared with a share of 26 and 8 % for

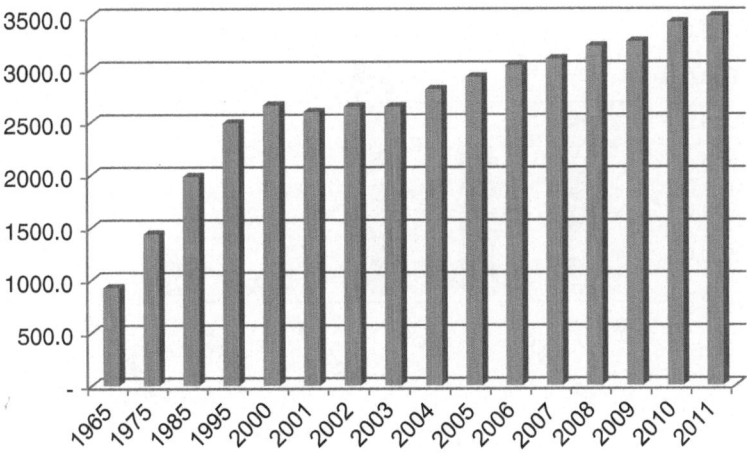

Fig. 3.1 Worldwide hydroelectricity consumption in TWh (Reproduced from BP (2012))

oil and natural gas respectively. Also, around 30 % of energy consumption in Sweden has been supplied by hydropower (66.5 TWh). China, Brazil, and Canada are the top three hydroelectricity producers worldwide with 694.0, 429.6, and 376.5 TWh generated respectively. Although hydroelectricity consumption in Norway and Sweden cannot be compared with these countries, they have significant hydropower generation, particularly relative to their size and total electricity supply. Figure 3.1 shows the general trend of worldwide hydroelectricity consumption from 1965 to 2011.

Total hydropower capacity is forecast to increase from 1,607 GW in 2011 to 1,680 GW in 2035 (IEA 2012b). According to the *WEO 2012* report, China is going to almost double its capacity, which will enable it to have 420 GW installed hydropower capacity in 2035. It is close to the figure for the entire OECD in 2011. According to IEA estimations, capacity will increase sharply in India and Brazil too. It is forecast that capacity will grow from 42 to 115 GW in India and from 89 to 130 GW in Brazil (IEA 2012b). Some regions such as Europe and North America where the hydropower sector is mature are going to modernize current plants and develop storage capacity instead of developing new traditional facilities (Martinot and Sawin 2012). An IEA survey indicates issues such as availability of funding, political and market risks, and local environmental concerns as barriers in the development of hydropower capacity in Africa. Figure 3.2 shows worldwide primary energy consumption by fuel in 2011, based on BP statistics (2012). Energy technology differs between the two groups of countries with respect to coal, nuclear, and hydropower. The difference is attributed to their technological capabilities.

There are three kinds of hydropower generation plants: run-of-the-river, where the power is generated by the flow of a river; reservoir, where power is generated by releasing stored water; and pumped storage, where stored water is backed to the reservoir in order to be pumped again. Small-scale hydropower stations are

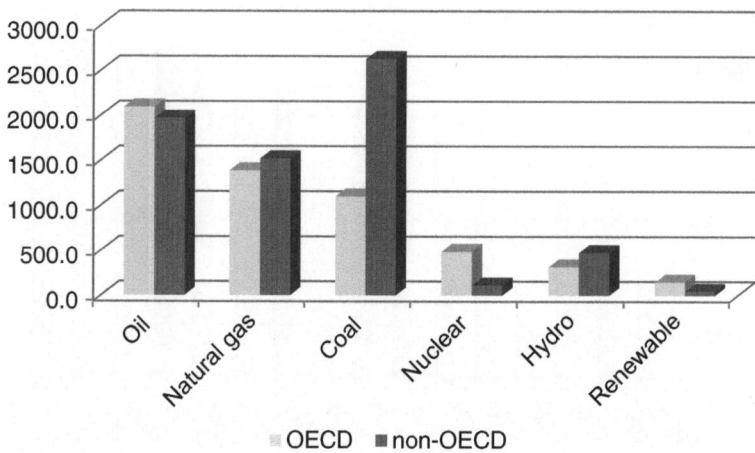

Fig. 3.2 Global primary energy consumption in 2011 (Reproduced from BP (2012))

run-of-the-river in most cases. Wirl (1989) examined conventional standards to evaluate hydropower plant projects and argued that the conventional cost–benefit analysis is not able to evaluate the expansion of plants appropriately. Wirl believed that the actual cost of hydropower plants is underestimated because of negative environmental externality and positive dynamic spillover effects.

Sinha (1993) estimated a model for a hypothetical site to simulate performance and economic aspects of combined wind/hydro/diesel power plants with pumped storage. His model comprised a wind energy conversion system, a mini/micro hydro plant, a diesel generator, and a pump. The results show that pumped storage does not have a significant effect when wind and water systems are applied but it could be used in sites without natural inflow. Gagnon and van de Vate (1997) discussed GHG emissions from hydropower plants and show hydropower is a good alternative compared to fossil fuel power plants in most cases. According to the results, a typical GHG emission factor is 15 g CO_2 equivalent/kWh, which is 30–60 times less than conventional fossil fuel power plants. Paish (2002) argued that the main advantages of small-scale hydropower include a more concentrated energy resource than wind or solar power, predictability, on-demand availability, limited maintenance, long-lasting technology, no fuel, and no environmental impact. He also pointed out certain shortcomings: site-specific technology, limitation of expansion activities, monsoon condition, conflicts with fisheries, and lack of knowledge to apply this technology in many areas.

Lehner et al. (2005) applied a model to analyze impacts of climate change on Europe's hydropower potential at a country scale. They analyzed gross hydropower potential and developed the potential of current plants in order to capture a real picture of present and future power generation. The results strongly indicated that the hydropower potential in Europe is influenced by climate change with a reduction of 25 % and more for southern and southeastern European countries. It is estimated that gross hydropower potential for Europe will decrease by about 6 % by the 2070s, while the reduction rate for developed hydropower potential will be 7–12 %. Lehner et al. (2005) stated that significant adoption is required for water management in the future and to support the necessity of developing mitigation strategies for the whole of Europe. Ehnberg and Bollen (2005) investigated the availability of a hybrid power plant when it is constituted by a combination of solar and small hydro installation. They used a small reservoir instead of a flow-of-river unit and assumed that hydro energy is not used during sunny periods. A model was simulated for four different combinations: solar power, solar power and storage, solar and hydropower, and solar and hydropower with storage. The results show that a combination of different sources should be hired in order to have a reliable supply. They also indicate that a combination of solar power and small reservoir is favorable compared to the other options.

Kaldellis et al. (2010) introduced a methodology to measure the size of pumped hydro storage (PHS) systems to take advantage of excess wind energy generated by local wind farms and rejected by the local power grid because of electrical limitation. Their findings show that the ability of the PHS system has made significant contributions in the electrification of remote islands. This methodology could be developed to apply for all hybrid projects constituted of a combination of wind farms, pumped storage, and hydroturbine. Kapsali and Kaldellis (2010) investigated the feasibility of a wind-based PHS system that is able to supply local power in an Aegean Sea island. The PHS systems located at isolated sites are able to use rejected wind energy produced by wind farms. The results showed that the project is viable in technical and economic terms. The share of renewable energy sources (RESs) will increase by 9 % after installation of the project, reaching about 20 % of local power consumption. It is indicated that the PHS project could be considered more environmentally friendly than conventional plants because it takes only required energy during the low demand period of the local grid when the thermal units operation generate less gas emissions.

Deane et al. (2010) reviewed current and planned pumped hydro energy storage (PHES) and analyzed technical and economic drivers for developing PHES. According to their results, current trends for developing PHES show an intention to enhance or build pump-back plants instead of pure pumped storage, which is partly due to lack of new feasible sites in economic terms. The capital cost for the proposed project in reviewed sites is estimated at 470–2,170 Euro/kWh. It is stated that developers of new PHES, particularly in Europe, intend to have hybrid wind–hydropower plants. Raadal et al. (2011) reviewed life-cycle GHG emissions from wind and hydroelectricity production compared to conventional fuels, nuclear energy, and other types of renewable energy sources. According to their results, GHG emissions produced by the run-of-the-river hydro plant life cycle analysis show the lowest variation among the examined technologies.

Yang and Jackson (2011) investigated the historical development of PHES in the USA and analyzed case studies, disputed projects, and challenges in the future development of these projects in the USA. Their findings show that interest in PHES systems has increased worldwide in recent years and it is expected that a capacity of

76 GW will be installed by 2014. The Federal Energy Regulatory Commission has granted 32 preliminary permits to 25 licensees in the USA to develop new PHES facilities. Yang and Jackson pointed that PHES development may be influenced by increased supply of unconventional natural gas and make it uncompetitive for using in the peak time of electricity networks. However, they argued, the possibility of a new law for price or to impose a limitation for carbon dioxide emissions could stimulate the economic outlook of PHES.

Connolly et al. (2011) applied a deterministic model to compare three operation strategies for optimizing profit in a PHES facility with a 360 MW pump, 300 MW turbines, and 2 GWh storage utilizing price arbitrage on 13 electricity spot markets. They found that an optimal strategy is achieved on day-ahead electricity prices and 97 % of profits could be earned by this strategy. They indicated that a long-term forecast is not required in order to maximize profit using electricity price arbitrage. Monteiro et al. (2013) estimated a short-term forecasting model for hourly average power generation of small hydro power plants (SHPPs). This model is constituted of three modules: estimation of daily average, final forecast of hourly average power generation, and dynamic adjustment by recent historical data. Monteiro and associates argued that a practical solution for technical and economic problems created by SHPPs is available due to this model. It is concluded that a power generation forecast is required to operate SHPPs appropriately for preparing bid offers in the markets and for maintaining the schedule of power plants.

A summary of the empirical research covering energy generated by hydropower and their findings is found in Table 3.1. The subjects studied include economic modeling, GHG emissions from hydropower, optimal size of plants, comparison with alternative renewable technologies, sector opportunities and barriers, and development forecasts. While hydropower is found promising in terms of reducing GHG emissions, PHE systems are efficient in electrification, but the market might be affected by the unconventional natural gas market development.

3.2.2 Wind Power Technologies

Wind power installed capacity has increased from 4.8 MW in 1995 to more than 239 GW in 2011. Today, wind turbines can generate as much electricity as conventional power plants. Wind energy has made its most significant contribution in China, the USA, and Germany, where cumulative installed capacities are 62, 47, and 29 GW respectively. Figure 3.3 shows the wind installation capacity trend worldwide based on BP (2012).

The trend shows that wind capacity installation has increased continuously during the last two decades. The IEA estimates that global capacity will increase from 238 GW in 2011 to almost 1,100 GW in 2035, 80 % of which will be constituted of onshore wind turbines (IEA 2012b). According to this report, offshore wind capacity is growing quickly and is expected to increase from 4 GW in 2011 to 175 GW by 2035 due to governmental support. This target will be achieved if required investment is made based on the designed plan. It is forecast that around

Table 3.1 Empirical research on hydropower

Authors	Subject	Result
Sinha (1993)	Modeling the economics of combined power systems	Pumped storage does not have significant effect when wind and water systems are applied. However, it could be used in sites without natural inflow.
Gagnon and van de Vate (1997)	GHG emissions from hydropower	A typical GHG emissions factor is 15 g CO_2 equivalent/kWh, 30–60 times less than fossil fuel generation.
Paish (2002)	Small hydropower technology	Main advantage: more concentrated energy, predictability, on-demand availability, limited maintenance, long-lasting technology, no fuel, and no environmental impact.
Lehner et al. (2005)	Impact of climate change on hydropower in Europe	Climate change makes a reduction by 25 % and more in hydropower potential for southern and southeastern Europe.
Ehnberg and Bollen (2005)	Reliability of a small power system with solar and hydro energy	A combination of different sources should be utilized to have reliability. Combination of solar power and small reservoir is more favorable.
Kaldellis et al. (2010)	Analysis of wind-based pumped hydro energy storage (PHES)	The ability of PHES has a significant contribution in electrification of remote islands.
Kapsali and Kaldellis (2010)	Combining hydro and variable wind power generation	PHS systems are viable in technical and economic terms at isolated sites.
Deane et al. (2010)	Techno-economic review of pumped hydro energy storage	Capital cost for PHES is estimated at 470–2170 Euro/kWh. It is intended to have hybrid wind–hydropower plants in Europe.
Raadal et al. (2011)	Life cycle GHG emissions from the generation of wind and hydro power	GHG emissions from wind and hydropower varies from 0.2 to 152 g CO_2-equivalents per kWh. Run-of-river hydro plant has the lowest.
Yang and Jackson (2011)	Opportunities and barriers to PHES in the US	PHES may be influenced by developing unconventional natural gas and make it uncompetitive. The possibility of a new law for price or emission could stimulate its outlook.
Monteiro et al. (2013)	Forecasting model for power production of small hydro power plants	Power generation forecast is required to operate small hydro power plants appropriately for preparing bid offers in the markets and for maintaining the schedule of power plants.

980 billion USD is required to invest during 2010–2020, which will increase to 1,634 and 3,307 billion USD for 2020–2030 and 2030–2050 respectively (IEA 2012b). Figure 3.4 shows the breakup of country groups for investment needs.

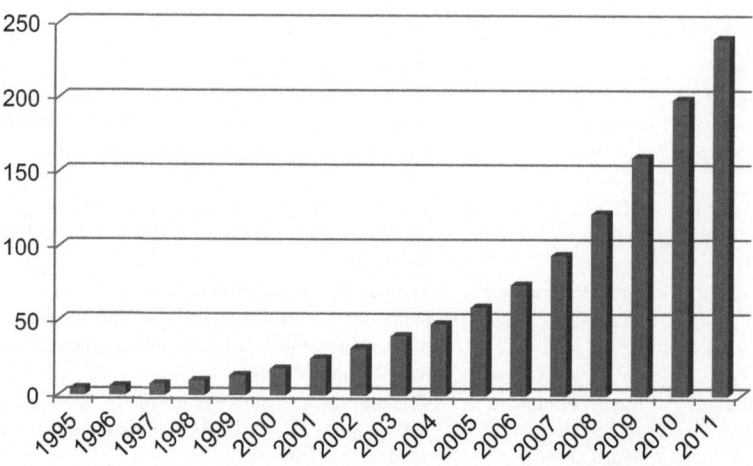

Fig. 3.3 Cumulative installed wind turbine capacity in GW (Reproduced from BP (2012))

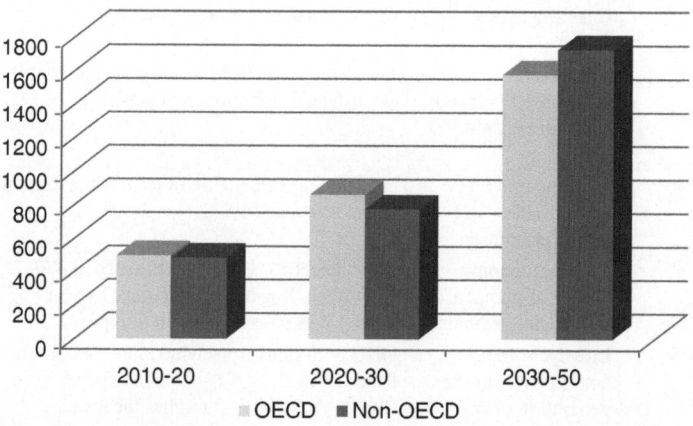

Fig. 3.4 Investment to achieve wind generation targets in billion USD (Reproduced from IEA (2012b))

Figure 3.4 shows that the OECD countries will fall behind the non-OECD countries by 2030–2050. The main portion of investment in the non-OECD group (almost 50 %) belongs to China.

"Of all solar (renewable) technologies, wind turbines are the closest to being economically competitive with fossil-fired systems" (Tester et al. 2005). Generally, renewable energy technologies are classified on the basis of the source of energy such as solar, wind, and biomass. But Tester et al. (2005) pointed that each renewable energy type comes from one of the three primary energy sources: solar radiation, gravitational forces, and heat generated by radioactive decay. They argued that solar, thermal, and photovoltaic energy are produced by capturing a fraction of incident solar energy. Wind, hydro, wave, ocean thermal, and biomass energy also are produced by solar energy indirectly. According to Tester et al. (2005), this competency could improve in the long term. Also, they estimated the lifetime-levelized cost for wind power as equal to 6.5 cents per kilowatt hour, which could be more compatible with natural gas combined cycle gas turbine (CCGT) and coal power plants if the externality costs such as SO_x, NO_x, and CO_2 emissions are considered. Furthermore, the authors added more advantages for wind power plants including installation as turnkey contracts within a short period, lower investment compared to nuclear and hydroelectric plants, economies of mass production free of fuel cost, and improvement in the operating and maintenance cost.

Ngô and Natowitz (2009) pointed out that there are some problems in using wind energy sources including intermittency of wind energy and additional cost for power transmission to residential areas because wind turbines are installed in windy sites where the population density is usually low. Offshore wind turbines are considered as an alternative to land-based turbines. Due to land limitation in some countries and probable opposition from local residents, offshore turbines could be used. The largest offshore wind farm has been installed in Denmark including 80 turbines with a power of 2 MW. Ngô and Natowitz mentioned that Denmark exports most of the power generated by wind turbines because there is no domestic demand at the time of production.

Gipe (1995) argued there are some crucial factors to use wind energy successfully. From a financial viewpoint, these factors include cost, revenue, and expected returns on investment. Other factors such as the national energy policy may be important, but financial targets could be managed by tax. Costs include both installation and operation expenditures. Revenue is calculated on the basis of wind resources, the turbine's performance, and the quantity of energy produced. For a wind plant, this value is defined by the purchase power rate or feed-in tariff. In view of a household living off grid, the value is calculated on the basis of the price of the electricity that they have to buy from the utility plus transmission cost to their house. The price of wind energy depends on how much energy is necessary for local residents. Therefore, feasibility and minimum required speed for wind turbine to be economical is related to the issue of how much wind energy is worth.

Gipe (1995) conducted two case studies in Europe and Great Britain and explained that wind energy is highly valued in northern Europe and 5.0–6.5 m/s is enough for wind speed to generate feasible energy, but the average speed in Great Britain should be over 7.0 m/s due to risks associated with tariffs. Gipe argued that wind turbines could be successful when there is a market for power generated. Some households may need to sell a part of wind turbines' excess power generation back to the local utility. An agreement is required between parties to process this transaction. At this step, feasibility depends on the government policy for pricing. In the USA, utilities are allowed to determine the price that is paid to individual generators, and it would be just a fraction of the retail price. Consequently, these individual producers may install smaller and less cost-effective wind turbines than a comparable producer in Denmark where they are allowed to sell surplus electricity at higher prices.

Devine (1977) used an input–output approach to calculate energy in a 1,500 kW (e) wind turbine used to displace fossil fuel in a power system. He compared five rations for delivered electricity and found this system is able to displace a part of the fossil fuel equivalent. Haack (1981) calculated the net energy of a small wind conversion system in the USA and compared it with other electricity sources such as coal-fired power plants, coal gasification gas-fired power plants, gas-fired power plants, nuclear power plants, and other generation sources. He estimated the amount of energy through a simulation model that takes into account wind speeds, residential electricity demands, and parameters from the generator, inverter, and storage components. The results show that net energy obtained by this wind system is better than other systems. Haack argued that the additional steps used in the process of obtaining fuels by new technologies decrease the efficiency of conversion. Also, the energy cost of materials that are used in different steps such as extraction, processing, or transportation of fuels will decrease, while the energy cost of wind generated is not expected to increase during the time.

Schleisner (2000) examined energy consumption and emissions generated through production and manufacturing of materials for onshore and offshore wind farms through a life-cycle analysis (LCA) model in Denmark. He calculated the weight of materials and the energy required for the production, manufacturing, and disposal processes. The emissions of N_2O, CH_4, and CO are converted to CO_2 equivalents by related factors. Schleisner (2000) also compared the primary energy used in production and disposal of materials in order to calculate the energy payback time. Based on his research, the energy payback time would be 0.39 years or less than 2 % of a 20-year lifetime if an estimation of 40 % is assumed for energy efficiency. Schleisner estimated the external cost of CO_2 emissions for wind farm on land and offshore as 0.8–1.2 mECU/kWh and 1.0–1.6 mECU/kWh respectively. The results show that damages for offshore wind turbines are larger due to the amount of materials used for the foundation and sea cables. To provide a comparative example, the external cost for a nuclear power plant in Germany has been estimated as 4.4–7.0 mECU/kW (Bodansky 2005).

Lenzen and Munksgaard (2002) conducted a review of energy and CO_2 life-cycle analysis of wind turbines. They found that small wind turbines of 1 kW require a considerably higher life-cycle energy than large-size turbines of 1 MW and argued that this deviation is due to values for the energy required for the materials, the analysis scope, the methodology used, the country of manufacturing, recycling component, and choice of concrete or steel for tower. Their research suggests that using a standardized methodology and input–output–based hybrid techniques can minimize uncertainties in the life-cycle assessment. Liberman (2003) employed the Monte Carlo simulation to analyze economic payback and life-cycle assessment of 11 modern, utility-scale wind turbines in the USA and used hourly meteorological data to evaluate 239 locations. The result shows wind turbines are not feasible at all locations but they could be superior to generators using natural gas or coal at locations with favorable wind sources.

Korpaas et al. (2003) used an algorithm to analyze the optimal energy exchange together with energy storage in the market for a certain period. Transmission constraints and the intermittent character of wind energy have been taken into account in this research. The results show that energy storage enables wind power plants' owners to take advantage of spot markets. Korpaas et al. (2003) argued that energy storage devices are an expensive alternative compared to developing power networks but they could be a feasible option for those places where grid expansions are not possible due to adverse environmental effects. Lenzen and Wachsmann (2004) conducted a life-cycle assessment to compare energy and CO_2 embodied in a particular wind turbine (E-40) with a nominal power of 500 or 600 kW manufactured in Germany and Brazil. Comparing economic structure and energy resources shows that the CO_2 balance is much lower in Brazil than Germany. This is because natural gas and nuclear power plants play a key role in Germany, while firewood and sugarcane-based alcohol are used exclusively in Brazil. Korpaas et al. investigated five scenarios for the production and operation of a particular wind turbine in these countries and found that CO_2 emission is considerably lower if the turbine is manufactured in Brazil. The results show that a production shift abroad could be a good solution to achieve emission reduction.

Wagner and Pick (2004) calculated energy yield ration and cumulative energy demand for two types of wind turbines (1.5 and 0.5 MW) in three sites— coastal, near coastal, and inland. Based on their results, energy payback time would be 3–7 months and energy yield ratio 38–70 depending on type and site. Wagner and Pick also found that deviation of the energy yield ration for different types is just 10 %. Klaassen et al. (2005) used the learning curve to examine how cost-reducing innovation is influenced by public R&D support for wind farms in Denmark, Germany, and the UK. They found that Denmark has developed small wind turbines as a result of R&D support and demonstration projects as well as investment subsidies. In Germany, R&D support to develop large-scale wind farms failed, but it was successful in the case of small wind turbines. In the UK, R&D support was insufficient with regard to the type of wind turbine, but the subsidy program was able to decrease the costs. Based on their results, Klaassen et al. (2005) estimated a rate of 5.4 % for learning-by-doing and 12.6 % for learning-by-research.

Benitez et al. (2008) used a nonlinear optimization program by load data for Alberta's grid in Canada to examine the economic and environmental effects of wind energy penetration in the power network and the extent to which hydropower storage is able to control wind energy intermittency. Based on their calculations, the generation cost of wind energy turbines is estimated at 37–68 USD/MWh, and the reduction cost of CO_2 emissions would be 41–56 USD/tone. The results show that hydropower could offset most of the peak load demand and eliminate building gas-fired generators for peak time. Tremeac and Meunier (2009) used life-cycle assessment to examine the environmental impact of 4.5 and 250 MW wind turbines taking into account all steps including manufacturing, transportation, installation, maintenance, disassembly, and disposal in their analysis. They found that wind energy could be the best environmental solution to mitigate climate change and supply electricity in off-grid areas if three conditions are considered: first, using highly efficient turbines in a proper site in terms of wind source; second, consuming less energy in the transportation stage; and third, correctly performing the recycling process.

Blanco (2009) investigated recent studies about wind energy manufacturers in order to categorize the generation costs for onshore and offshore turbines. She also analyzed the supply chain constraints and main factors that caused a cost increase of 20 % during the last 3 years. Based on her results, generation cost is estimated at 4.5–8.7 Eurocents/kWh for onshore and 6.0–11.1 Eurocents/kWh for offshore wind turbines. The rising price of key raw materials and the unexpected increase in demand for wind turbines have been the main reasons for the 20 % increase in generation cost. Blanco believed that an appropriate, stable, and long-term policy framework is required to decrease the generation costs of wind energy in the long run. She argued these policies could focus on R&D in optimization of size of turbines and new materials for blades, remote-control devices for operation and management, advanced forecasting techniques, and availability of sufficient funds.

Crawford (2009) used a hybrid embodied energy analysis approach to assess life-cycle energy and GHG emissions for 850 kW and 3.0 MW wind turbines and examine the impact of turbine size on energy yield ratio. He argued that methodologies used in previous research on life-cycle energy are incomplete due to some limitations and errors in the quantification of inputs to product and valuation of energy requirements for supporting goods and services. His results estimated energy yield ratios of 21 and 23 for small- and large-scale wind turbines. Crawford found that the size of wind turbines is not an important parameter in optimizing life-cycle energy performance. Kubiszewski et al. (2010) reviewed the literature on the net energy return for wind turbine power generation, examining 119 wind turbines from 50 different research studies published during 1977–2007. The results show that average energy return on investment (EROI) for all studies including operational and conceptual is 25.2, while it is 19.2 for operational studies which place wind in a good position compared to fossil fuels, nuclear power, and solar power generation.

Sundararagavan and Baker (2012) applied a cost analysis for different types of energy storage technologies that are useful in mitigating the uncertainty of integrating wind turbines and power grids due to the intermittency of wind power. They argued three key factors are required for this integration—load shifting, frequency support at transmission and distribution levels, and power quality to smooth power fluctuations. The results show that no single technology could dominate all these three applications. The authors believed that assumptions about interest rates play a crucial role in making a difference between technologies and, importantly, selecting good technology depends on the perspective of decision-makers.

A summary of the empirical research on energy generated by wind power and their findings is found in Table 3.2. The subject of research is mainly concentrated on the life-cycle assessment of wind power. The results suggest variations in the energy payback time.

3.2.3 Solar Power Technologies

During the last few decades, researchers have investigated the economic feasibility of solar power for residential, commercial, and industrial consumption. Industrial

Table 3.2 Empirical research on power generated by wind

Authors	Subject	Result
Haack (1981)	Net energy analysis of small wind energy conversion systems	Small wind electric systems are energetically competitive and at an advantage over other electricity generating systems.
Schleisner (2000)	Life-cycle assessment of a wind farm and its externalities	Energy payback time (EPBT) would be 0.39 years or less than 2 % of a 20-year lifetime if an estimation of 40 % is assumed for energy efficiency.
Lenzen and Munksgaard (2002)	Review of energy and CO_2 life-cycle analysis of wind turbines	Minimizing uncertainties in life-cycle assessment is suggested by using a standardized methodology and input–output-based hybrid techniques.
Liberman (2003)	Economic payback and life-cycle assessment of utility-scale wind turbines in the US	Wind turbines are not feasible at all locations, but they could be superior to generators using natural gas or coal at proper locations.
Korpaas et al. (2003)	Operation and sizing of energy storage for wind power plants	Energy storage enables wind power plants' owners to take advantage of spot markets. They are expensive, but it could be a feasible option.
Lenzen and Wachsmann (2004)	Geographical variability in life-cycle assessment	A production shift abroad could be a good solution in order to achieve emission reduction.
Wagner and Pick (2004)	Energy yield ration for wind energy converters	Energy payback time (EPBT) would be 3–7 months and energy yield ratio is 38–70, depending on type and site.
Klaassen et al. (2005)	The impact of R&D on innovation for wind energy in Denmark, Germany, and the UK	A rate of 5.4 % is estimated for learning-by-doing and 12.6 % for learning-by-research to develop wind farms.
Benitez et al. (2008)	The economics of wind power with energy storage for Alberta in Canada	Generation cost of wind energy turbines is estimated as 37–68 USD/MWh and reduction cost of CO_2 emissions would be 41–56 USD/ton.
Tremeac and Meunier (2009)	Life-cycle analysis of 4.5 and 250 MW wind turbines	Wind energy could be the best environmental solution to mitigate climate change and supply electricity in off-grid areas.
Blanco (2009)	The economics of wind energy	Generation cost is estimated as 4.5–8.7 Eurocents/kWh for onshore and 6.0–11.1 Eurocents/kWh for offshore wind turbines.
Crawford (2009)	Life-cycle analysis and GHG emission analysis for wind turbine	The size of wind turbines is not an important parameter to optimize life-cycle energy performance.
Kubiszewski et al. (2010)	Net energy return for wind power systems	Average energy return on investment (EROI) for all studies including operational and conceptual is 25.2, while it is 19.2 for operational studies.
Sundararagavan and Baker (2012)	Evaluating energy storage technologies for wind power	Assumptions about interest rates play a crucial role in making a difference between technologies, and selecting a good technology depends on the perspective of decision makers.

countries like Japan and Germany are looking for alternative sources of energy
such as solar power due to the limited availability of natural primary energy
sources. In the early 1990s, Japan started taking advantage of large-scale electricity
generation by solar photovoltaics, followed by Germany. Currently, both countries,
with multibillion-dollar industries in solar power, have taken the lead in the
manufacturing and production of solar power technologies. In view of its industrial
requirements, China has developed an extensive solar power capacity and has
decreased the cost of solar power generation by taking advantage of cheap labor
and government subsidies.

Alongside cost reduction of power generated through conventional solar PV
technologies, advancement and high efficiency in concentrated solar power technologies in the USA has resulted in more reduction in the cost of electricity in solar
power industry (Gevorkian 2012). On the other hand, there are some negative effects
caused by solar technologies, including visual impact on buildings' aesthetics,
routine and accidental release of chemicals, use of land, impact of large PV systems
on ecosystems, and construction activities for solar thermal energy (Tsoutsos et al.
2005). The solar PV market has experienced an extraordinary growth during the last
5 years. The market increased from 9,564 MW in 2007 to 69,371 MW in 2011.
Figure 3.5 shows its trend during1996–2011 based on BP (2012).

Almost 30 GW of new capacity was installed worldwide in 2011, leading to an
increased total world capacity of 69 GW. A major part of this new capacity surged
at the end of the year due to tariff support policies, the impending expiration date
of some policies, and price reductions. Turkey increased its capacity by 1,353 % in
2011 over 2010. Bulgaria, Italy, Slovakia, and Greece also increased their capacity
more than threefold. It is forecast that PV industrial mass production will be
established worldwide during 2010–2020 and PV systems will be integrated with
the power grid post 2020. Figure 3.6 shows the investment needs to install solar PV
systems by 2050 (IEA 2012a)

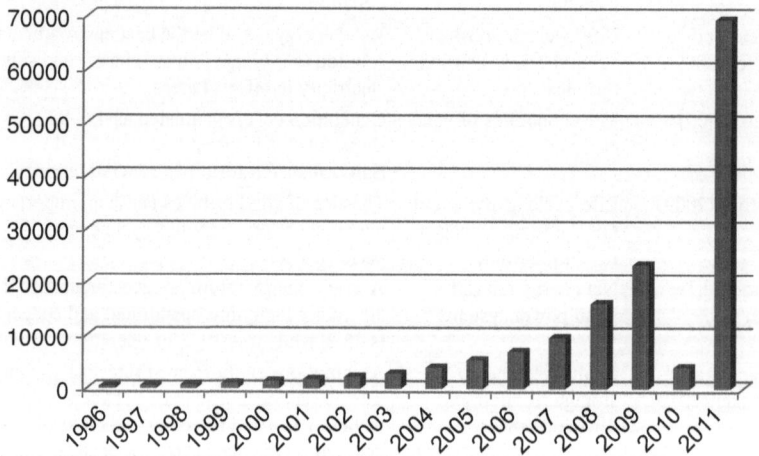

Fig. 3.5 Cumulative installed solar PV capacity in MW (Reproduced from BP (2012))

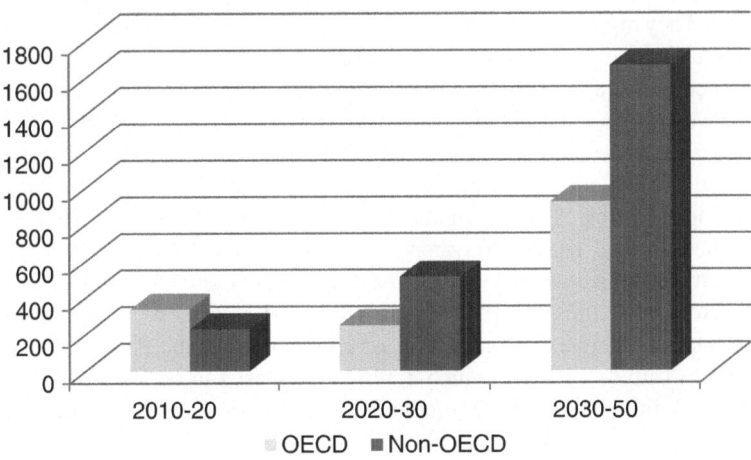

Fig. 3.6 Investment to achieve solar PV power generation targets in USD billion (Reproduced from IEA (2012a))

Similar to wind energy, solar energy depends on weather conditions. Variations in the weather such as clouds and pollution could affect solar power generation. There is a major difference between wind and solar power because solar energy is available just during daylight hours. Therefore, solar power generation varies by season, location, and time of the day. Many technologies are being used to deploy solar radiation including thermal solar energy, concentrated solar power plant (CSP), solar chimneys or towers, and photovoltaic systems (Ngô and Natowitz 2009). An advantage of photovoltaic technology compared to other technologies is the opportunity to integrate a PV collector into the building by turning external walls, windows, and roofs into a PV collector. However, there could be some environmental and health concerns because of the materials used for PV systems (Tester et al. 2005). Furthermore, PV performance could be reduced because of dust effects on glass and transparent materials. Sarver et al. (2013) examined and summarized the research and challenges regarding dust problems for solar panels.

Gordon (1987) analyzed optimal sizing of stand-alone photovoltaic power generation systems to design a cost-effective alternative for conventional fossil fuel generators in developing countries where most people live in rural and off-grid areas. Frankl et al. (1997) evaluated benefits of building integrated PV systems compared to conventional PV power plants through life-cycle analysis, maximizing energy efficiency, and CO_2 reduction potential. The results show favorable effects for building integrated PV systems in terms of energy production and CO_2 reduction emissions. The study estimated a CO_2 yield of 2.6 and 5.4 for conventional PV power plants and building integrated systems respectively. It also forecast that these benefits will increase in the future by advanced PV technologies.

Market interests to deploy renewable energy including solar power are increasing globally. Oliver and Jackson (1999) proposed some markets as the main markets for solar PVs. They argued that satellites, remote industrial areas, remote communities, solar home systems, remote houses, and consumer products (indoor applications) could be considered as a niche market for solar PV power. Nieuwenhout et al. (2001) investigated experimental evidences for solar home systems (SHSs) in developing countries and found that an adequate level of service infrastructure is required for the project viability of solar PVs. The results show that a few problems such as lack of information on user experience, possible negative impacts of subsidies, limited choice of system size, and insufficient market transparency present difficulties. Kolhe et al. (2002) analyzed the economic feasibility of a stand-alone solar photovoltaic (SAPV) system compared to a diesel power plant as a conventional alternative for India. The results show that the PV system has the lowest cost of up to 15 kWh, which could be increased to 68 kWh/day when the economic conditions are more favorable. Kolhe et al. (2002) argued that the break-even point will go up if the cost of PV decreases and diesel costs increase.

Jäger-Waldau (2006) examined the European photovoltaic market and pointed out that given a growth rate of 40 % in 5 years, photovoltaic is one of the fastest-growing industries. He argued this industry needs a reliable political framework to ensure returns on investment and continuous research to find cost-effective material, device designing, and increasing efficiency. Nawaz and Tiwari (2006) analyzed energy payback time and CO_2 emissions of PV systems in India. They estimated embodied energy to produce PV modules at macro and micro levels with an assumed irradiation of 800–1,200 W/m^2 in different sites and concluded that energy payback time (EPBT) depends on solar radiation, efficiency of the PV system, and balance of the system. It is estimated that EPBT is in the range of 7–26 years and CO_2 emission reduction by existing technology is expected to be in the range of 18–160 kg/m^2/year.

Shum and Watanabe (2007) compared the use of PV technology in Japan and the USA and applied two models named manufactured technology and information technology to explain differences in strategies. Ito et al. (2008) examined five types of 100 MW very large-scale photovoltaic (VLS-PV) power generation in economic and environmental terms. They compared five types of PV modules installed in the Gobi desert (China) and investigated three studies about interest ratios, transmission distances, and ambient temperatures. The results showed that energy payback time is 1.5–2.5 years and the CO_2 emission rate is 9–16 g/kWh. Also, the generation cost was estimated at 11–12 US Cent/kWh for using 2 USD/W PV modules and 19–20 US Cent/kWh for using 4 USD/W PV module price.

Fthenakis and Kim (2007) considered the entire life-cycle energy for solar and nuclear power generation to compare their potential for GHG emission reduction in the USA. They used data from 12 photovoltaic companies and reviewed nuclear fuel life cycles in the USA, Europe, and Japan. The results showed that GHG emissions (based on CO_2 equivalent) are 22–49 g/kWh (average USA) and 17–39 g/kWh (southwest) for solar energy and 16–55 g/kWh for nuclear power. In another study, Fthenakis et al. (2008) analyzed life-cycle GHGs, criteria pollutants, and heavy metal emissions for four types of PV technologies including multicrystalline silicon, monocrystalline silicon, ribbon silicon, and thin-film cadmium telluride. They found that thin-film cadmium telluride has the least amount of emissions among the four types. The differences in emissions for various PV technologies are too small compared to the conventional energy that is to be replaced with PV systems. Feltrin and Freundlich (2008) examined different solar PV technologies ranging from silicon to thin films and concentrated systems based on global available material reserves for large-scale power generation at terawatt-level deployment of photovoltaic energy. According to their findings, in spite of adequate availability of silicon, crystalline Si-based solar cells could not reach the terawatt level easily in a large scale-up of technology. Therefore, improvement and innovation are required to overcome the material challenge.

Raugei and Frankl (2009) proposed three alternative scenarios for the future development of PV systems up to 2050 and argued these are likely to play an important role in the future energy mix if appropriate economic incentives are continued for the next two decades. Fthenakis et al. (2009) used hourly load data for the entire USA and 45-year solar irradiation data from the southwest region and proposed a plan based on PV and CSP technologies, integrated with compressed air energy storage (CAES) for PV and thermal storage for CSP. They believed that solar energy has been a minor contributor so far due to the costs involved and the intermittent character of solar energy. However, cost reductions made by new emerging technologies enable solar power to be compatible with fossil fuels. They show that solar power has the capability to supply 69 % of the total electricity demand and 35 % of the total energy demand in the USA by 2050. Based on their research, the figure could be increased to 90 % if the scenario is extended to 2100.

In a recent study, Huo et al. (2011) applied the Granger causality relationship between PV market sale and manufacturing development in the USA, Germany, China, and Japan. The results show that the growth of market sale affects the innovation scale in the USA, Germany, and Japan. Also, there is a bidirectional relationship between PV market sale and manufacturing development in the USA and Germany. Huo et al. (2011) argued that the manufacturing sector could influence the dynamics of market sale. Lin (2011) investigated key development factors that create competitiveness of the solar PV industry in Taiwan and the causal effects of these factors. It is indicated that local demand conditions, government support, and related supporting industries are three factors that influence the solar PV industry strongly. Branker et al. (2011) calculated levelized cost of electricity (LCOE) generation of solar PV for a case study in Canada. They examined the grid parity of the PV system compared to electricity price of conventional technology to analyze cost effectiveness and found that solar PV has already met grid parity in some locations due to cost reduction. It is expected that feasibility of the solar PV system will increase in the future as it expands geographically.

A summary of the empirical research covering the energy generated by solar power and their findings is provided in Table 3.3. As with wind power, the main subject of research is life-cycle and market analysis of PV systems and GHG emissions as well as reliability, supply capability, technology innovation, grid parity, and economic incentives.

Table 3.3 Empirical research on power generated by solar power

Authors	Subject	Result
Frankl et al. (1997)	Life-cycle analysis of PV systems in buildings	They estimated CO_2 yield of 2.6 and 5.4 for conventional PV power plants and building integrated systems.
Oliver and Jackson (1999)	Market for photovoltaic	Satellites, remote industrial, remote communities, solar home systems, remote houses, and consumer products could be considered as a niche market for solar PV.
Nieuwenhout et al. (2001)	Experience with solar home systems in developing countries	Lack of user experience, possible negative impacts of subsidies, limited choice of size, and insufficient market transparency appear to present difficulties.
Kolhe et al. (2002)	Economic feasibility of stand-alone solar PV compared with diesel in India	PV system has the lowest cost up to 15 kWh and it could be increased to 68 kWh/day. The breakeven point will go up if the cost of PV decreases and diesel costs increase.
Jäger-Waldau (2006)	European PV in worldwide comparison	Reliable political framework to ensure returns on investment and continuous research to find cost-effective material, device designing, and increasing efficiency are required.
Nawaz and Tiwari (2006)	Energy analysis of PV based on macro and micro levels in India	It is estimated that EPBT is in the range of 7–26 years and CO_2 emissions reduction by existing technology are calculated in the range of 18–160 kg/m^2/year.
Fthenakis and Kim (2007)	GHG emissions from solar and nuclear power	GHG emissions (based on CO_2 equivalent) are 22–49 g/kWh (average US) and 17–39 g/kWh (southwest) for solar energy and 16–55 g/kWh for nuclear power.
Ito et al. (2008)	Comparative study on cost and life-cycle analysis for very large scale PV	EPBT is 1.5–2.5 years and CO_2 emissions rate is 9–16 g/kWh. Also, the generation cost was estimated as 11–12 US Cent/kWh for using 2 USD/W PV modules and 19–20 US Cent/kWh for using 4 USD/W PV module price.
Fthenakis et al. (2008)	Emissions from PV life cycle	Thin-film cadmium telluride has the least amount of emissions among the four types of technology. The differences in emissions for various PV technologies are too small.
Feltrin and Freundlich (2008)	Material consideration for terawatt level deployment of PV	In spite of enough availability of silicon, crystalline Si-based solar cells could not reach the terawatt level easily in a large scale-up of technology.

(continued)

Table 3.3 (continued)

Authors	Subject	Result
Raugei and Frankl (2009)	Life-cycle impacts and costs of PV systems	If economic incentives are continued adequately for the next two decades, PV systems likely play a significant role in the future energy mix.
Fthenakis et al. (2009)	Feasibility for solar energy to supply the energy needs of the US	Solar power has capability to supply 69 % of total electricity demand and 35 % of total energy demand in the US by 2050. It could be increased to 90 % if the time period is extended to 2100.
Huo et al. (2011)	Relationship between PV market and its manufacturing	Growth of market sale affects innovation scale in the US, Germany, and Japan. Feasibility of solar PV system will increase in the future as it expands geographically.
Branker et al. (2011)	Solar PV levelized cost of electricity	Solar PV has already met grid parity in some locations due to cost reduction. Feasibility of the solar PV system will increase in the future as it expands geographically.

3.2.4 Geothermal Power Technologies

Geothermal is a type of thermal energy generated and stored in the Earth. It has been used for bathing, heating, and cooking for a long time. Geothermal energy is created by radioactive decay with the main temperature reaching 4,000 °C at the core of the Earth. While geothermal energy is available worldwide, there is an important factor called geothermal gradient to indicate a region as a favored place for deploying such energy. It measures the rate of increasing temperature when the depth in the Earth is increasing. For example, the geothermal gradient average in France is 4 °C/100 m with a broad variation from 10 °C/100 m in the Alsace region to 2 °C/100 m in the Pyrenees. In Iceland and volcanic regions, 30 °C/100 m may be reached (Ngô and Natowitz 2009).

Geothermal gradient is not the only dominant factor to measure the accessibility of geothermal energy. Permeability of the rocks, which determines the rate of flowing heat to the surface, is considered another important measurement in deploying geothermal energy. Geothermal has a big advantage compared to wind and solar energy—availability of 24 h through the year. According to Ngô and Natowitz (2009), an annual percentage of 80–90 could be reached. They estimated CO_2 emission produced by geothermal resources is 55 g/kWh, employing data from a survey of 73 % of the geothermal power plants. This value may be decreased to zero if geothermal fluid is reinjected into the ground. A total of 24 countries are using geothermal power plants now. Total installed capacity was 11 GW in 2011. Figure 3.7 shows cumulative installed geothermal power capacity worldwide based on BP (2012).

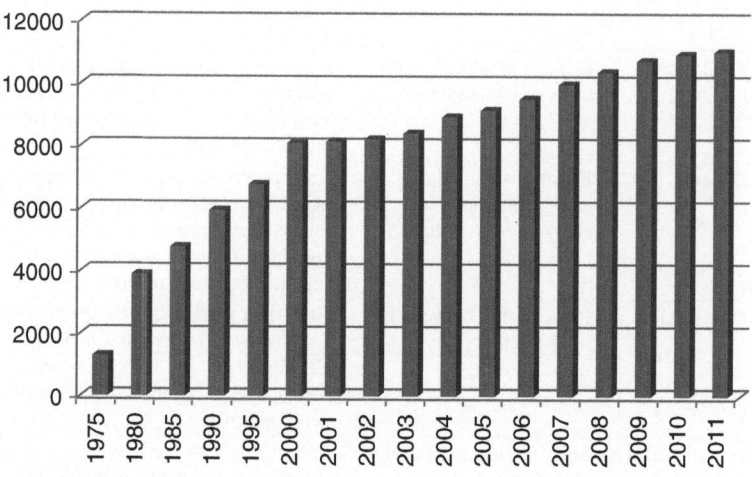

Fig. 3.7 Cumulative installed geothermal capacity in MW (Reproduced from BP (2012))

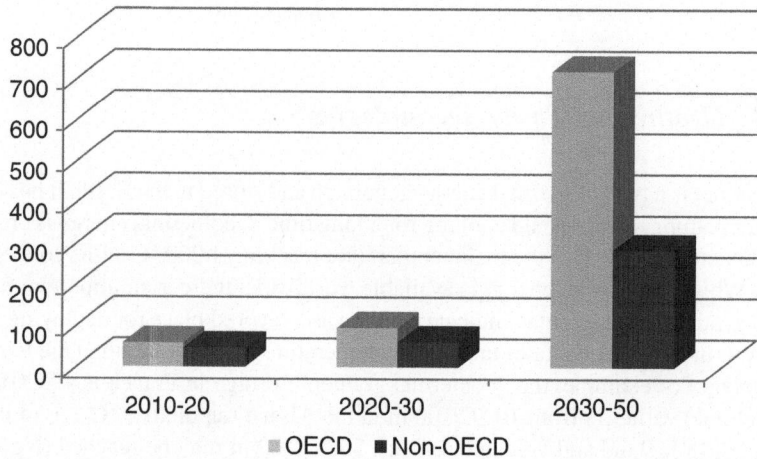

Fig. 3.8 Investment to achieve geothermal generation targets in billion USD (Reproduced from IEA (2012a))

Costa Rica, Turkey, and Iceland increased their capacity in 2011 over 2010 by 25.3, 21.2, and 15.7 % respectively. The majority of worldwide capacity (GW) has been installed in eight countries: the United States (3.1), the Philippines (almost 2.0), Indonesia (1.2), Mexico (0.9), Italy (0.9), New Zealand (0.8), Iceland (0.7), and Japan (0.5) (BP 2012). Geothermal capacity is expected to deploy coproduction geothermal power by exploiting water from oil and gas wells throughout 2015. Figure 3.8 shows the investment needs for geothermal development by 2050 for country groups based on IEA estimation (IEA 2012a).

Fridleifsson and Freeston (1994) investigated geothermal energy development based on worldwide experiences and forecast an estimate of 15–20 billion dollars as the total investment cost of geothermal in the world during the next decade. However, this estimation increased fivefold in the IEA report, which declared 104 billion dollars as the worldwide required investment in 2010–2020 to meet desired targets (IEA 2012a). Fridleifsoon and Freeston pointed that considering geothermal independence from weather conditions and storage capability, it could be used both for base load and peak power plants. They stated that environmental problems created by the release of steam, gases, and hot water to the rivers could be reduced by advanced technologies. They also argued that many countries, particularly developing countries, try to deploy geothermal energy with available technology but there are some difficulties due to lack of finance and knowledge infrastructure for training technicians. They proposed more training centers should be established through international programs.

Murphy and Niitsuma (1999) mentioned strategies for compensating the higher costs of geothermal electricity with environmental benefits and suggested some fiscal policy measures such as carbon tax to support and monetize the advantages of geothermal. They pointed Japan, Indonesia, and the Philippines as countries in which geothermal growth has remained high due to supportive policies performed by their governments. Stefansson (2002) applied the statistics methodology and employed data from Iceland to estimate investment costs required to build a power plant for unknown geothermal fields. He believed that price levels in Iceland are similar to Europe and the USA; therefore, the results could be applied to the other countries. Based on his findings, the total investment cost of geothermal power plants in the range 20–60 MW is estimated as 1,267 USD/kW in a known field and 1,440 USD/kW in an unknown field.

Lund et al. (2005) reviewed worldwide direct application of geothermal energy by employing data from 72 countries. The results showed that using 273,372 TJ/year of energy consumption in 2005 and assuming 6.06×10^9 J for each barrel of fuel oil, the estimated energy saving could be 128.9 million barrels of oil or 19.2 million tons of oil per annum. Lund et al. argued there will be an additional saving of 41.2 million barrels (6.2 million tons) of fuel oil or 7 million tons of carbon dioxide emissions if saving in the cooling mode of geothermal heat pump is considered.

Frick et al. (2010) applied a life-cycle analysis method on geothermal power generation from EGS (enhanced geothermal system) low-temperature reservoirs in order to show how much of the environment is affected by geological conditions. They found that geothermal binary power plants could not be explained by environmental indices because of different geological conditions. The results show that a life-cycle geothermal binary power plant is determined by materials and energy inputs. Therefore, successful access to reservoirs by minimum drilling is an important factor for low environmental impacts. The study found that less favorable geothermal heat and power generation could have a contribution in the energy system to enhance sustainability. In a related study, Saner et al. (2010) discussed energy consumption and GHG emissions and applied life-cycle assessment to examine environmental impacts due to ground source heat pumps (GSHPs) installed for deploying geothermal energy. The results classified the environmental impacts of the GSHP system into resource depletion (34 %), human health (43 %), and ecosystem quality (23 %). Also, the CO_2 emission equivalent has been estimated at an average of 63 tons for a life cycle of 20 years, that is, 31–88 % emission saving for Europe compared to conventional heating systems.

Purkus and Barth (2011) analyzed the German geothermal industry and emphasized the importance of political support and framework conditions in the electricity market. They argued that high investment costs and the risk of insufficient heat are considered as disadvantages of geothermal technology, but the core advantage of nonintermittency compared to other kinds of renewable energy sources could enable it to be considered as a reliable supply of base load power. Kaya et al. (2011) investigated a worldwide experience of reinjection in geothermal fields, employing data from 91 geothermal power plants, and found that a reinjection plan is required in order to reduce the risk of groundwater contamination. The results show that response of the geothermal reservoir to different strategies for reinjection depends on the geothermal system. Also, the authors stated that reinjection is an environmentally friendly method of wastewater disposal and should be performed properly because the consequent effects may be irreversible.

In a recent study, Chamorro et al. (2012) reviewed the worldwide status of geothermal energy and found that high-temperature technologies of flash and dry steam are the most developed geothermal power generation technologies. They defined four model plants— 1FMP (single flash model plant), 2FMP (double flash model plant), 3FMP (triple flash model plant), and DSMP (dry steam model plant)— to analyze geothermal systems. The results show that DSMP has the highest NPV amount and IRR factor with 1,013.6 M$ and 22.8 %. Also, the cost of electricity is 29.38 $/MWh for DSMP, which is estimated as minimum among the different models.

A summary of the empirical research covering the power generated by geothermal and their findings is found in Table 3.4. The main subject focus is on R&D, cost strategies, investment costs, life-cycle assessment, energy, and environment aspects of geothermal. Estimations of economic costs and life cycles of the technology are presented.

3.2.5   Other Renewable Power Technologies

There are some other types of renewable energy sources including biomass, ocean waves, and tides. Biomass is defined as living plant and organic waste, which is made by plants, humans, marine life, and animals. Based on Tester et al. (2005), the main advantage of biomass is availability—it is found widely in all places. Many

Table 3.4 Empirical research on power generated by geothermal

Authors	Subject	Result
Fridleifsson and Freeston (1994)	Geothermal energy R&D	It forecasted an estimate of 15–20 billion dollars for total investment of geothermal in the world during the next decade. Geothermal energy could be used for both base load and peak power plants.
Murphy and Niitsuma (1999)	Strategies for compensating higher costs of geothermal	They suggested some fiscal policy measures such as carbon tax to support and monetize the advantages of geothermal electricity.
Stefansson (2002)	Investment costs of geothermal power plants	Total investment cost of geothermal power plants in the range 20–60 MW is estimated 1,267 USD/kW in a known field and 1,440 USD/kW in an unknown field.
Lund et al. (2005)	Application of geothermal energy	Using 28,268 MWt installed capacity in 2005, estimated energy saving could be 128.9 million barrels of oil or 19.2 million tons of oil per annum.
Frick et al. (2010)	Life-cycle assessment of geothermal binary power plants	A life-cycle geothermal binary power plant is determined by materials and energy inputs. Successful access to reservoir by minimum drilling is an important factor.
Saner et al. (2010)	Life-cycle perspectives on geothermal systems	CO_2 emission equivalent has been estimated at an average of 63 tons for a life cycle of 20 years, that is, 31–88 % emission saving for Europe compared to conventional systems.
Purkus and Barth (2011)	Geothermal power production in future electricity markets	High investment costs and risk of insufficient heat are disadvantages of this technology. Non-intermittency could enable it to be considered as a reliable supply of base load power.
Kaya et al. (2011)	Reinjection in geothermal fields	A reinjection plan is required in order to reduce the risk of groundwater contamination. It is an environmental friendly method of wastewater disposal.
Chamorro et al. (2012)	Energy, environmental, and economic study of geothermal technology	Dry steam model plant (DSMP) has the highest NPV amount and IRR factor with 1,013.6 M$ and 22.8 %. Also, cost of electricity is 29.38 $/MWh for DSMP, which is estimated as minimum among the different models.

kinds of energy can be made from biomass: electricity, cooking, chemical feedstock, and so on. As a feedstock, biomass has lower sulfur content than coal. Therefore, lower emission is produced by combustion. In early 2000, the USA had an installed capacity of 11 GW from biomass including forest product and agricultural industry (7.5 GW), municipal and solid waste (3 GW), and other sources (0.5 GW) (Ngô and Natowitz 2009).

Deploying energy from oceans is considered as an interesting option because of the wide availability of ocean sources. There are six different resources that are available from oceans: offshore wind energy, wave power, marine current energy, ocean thermal energy conversion, tidal power, and osmotic power. The Bay of Fundy has the largest tidal range in the world, which enables it to support a power station with a capacity of 2 GW or more (Tester et al. 2005). We considered hydro, wind, solar, and geothermal energy because of their main contribution in power generated by renewable energy sources.

3.3 Energy Efficiency Technologies

The second of the two main solutions to reduce CO_2 emissions and overcome the climate change problem, besides the use of renewable energy, is the enhancement of energy efficiency. We have discussed the state of the art in technical feasibility and economic viability of deploying renewable energy sources and the possibility of substitution in a previous section. In this section, we are talking about energy efficiency technologies. Energy efficiency for electricity networks could be considered in different stages including power generation (here, the focus is on small size), transmission, distribution, and consumption. For this purpose, different technologies are available such as batteries, combined heat and power (CHP), virtual power plants (VPPs), and smart grids.

3.3.1 Electric Vehicles

A way to counteract the fluctuations that come with the irregularity of renewable energy production is the use of batteries. With the use of batteries, stable electricity can be produced on demand. However, as the cost of batteries is very high, the cost of renewable energy would increase even further. Therefore, a joint use of batteries could be a valid opportunity. In this case, the cost accounted to renewable energy would be less, and due to the higher production quantity of batteries, the unit cost would go down as well. An example of a joint use could be the use of batteries and electric vehicles.

Electric vehicles have the potential to be considered for electricity storage. Considering that the transportation sector is one of the main sources of emissions, improving fuel efficiency enables the largest fuel saving and CO_2 reduction in the short term. Thus, implementation of EVs and increasing their share in vehicle fleets can play a key role in the long term. IEA forecast (IEA 2012a) that an increased share of plug-in hybrid electric vehicles (PHEVs) is becoming more important for the next two decades and it will increase by up to 50 % by 2050 in the ETP 2012 2DS scenario (2DS scenario is consistent with the 450 scenario through 2035). Together with smart grid technology, EV can be used as electricity storage devices and a source of electricity if required (IEA 2012b).

Ford (1995) examined the impact of large-scale use of electric vehicles in southern California and analyzed that the Southern California Edison company is able to accommodate a large number of EVs with existing capacity, particularly if the charging system could be managed by smart control. Ford argued that EVs could improve load management, enhance efficiency, and save energy. He also calculated that EVs are able to reduce emissions valued about 9,000 USD per vehicle. Kempton and Letendre (1997) calculated present value costs for EV owners and benefits to utility and found that it is highly profitable for utility. Based on their results, all three vehicle/battery combinations are cost-effective power sources in peak time for the near term. They argued that if a part of the transportation sector is utilized for electric vehicles and connected to the electricity network, there will be less demand for base load generation and using intermittent renewable energy sources will become more applicable due to lack of concerns about time-of-day match between demand and supply.

Kempton and Tomić (2005) investigated the systems and procedures required to use energy in vehicles and the implementation of vehicle-to-grid (V2G) technology. The most important role of V2G could be to support renewable energy in emerging power markets through managing load and supply fluctuations made by the intermittency of renewable energy sources such as photovoltaic and wind turbines. They argued that after initially tapping EVs due to high value, market saturation, and cost reduction, the V2G fleet could be used as storage capacity for renewable energy power generation. Tomić and Kempton (2007) examined the economic feasibility of two battery-electric vehicles to supply power for a particular market in four US regulation service markets. The results show that V2G electricity is able to provide a significant income flow, helping feasibility of grid-connected vehicles and further supporting adoption. Lund and Kempton (2008) evaluated integration of renewable energy into the transport and electricity sectors by V2G. They applied a model to analyze energy integration for electricity, transport, and heating. V2G technology is able to provide storage for matching time of generation with time of load. Lund and Kempton found that adding EVs and V2G to power networks enables systems to be integrated with higher levels of wind electricity without extra power generation, making a huge reduction in CO_2 emissions.

Steenhof and McInnis (2008) analyzed three scenarios—the impact of increasing ethanol 85, hydrogen, and electricity-powered vehicles in the passenger transportation fleet, which was implemented in 2010 and is expected to reach 100 % of the new market by 2050. The results show that CO_2 emissions will be reduced from 153 Mt for electric vehicles to 156 Mt for hydrogen fuel cell vehicles by 2050. It is forecast that ethanol-driving cars will be cellulose based by 2050, generating a significant reduction in CO_2 emissions, but an unsustainable amount of crop residues will be required.

Andersen et al. (2009) introduced an intelligent electric recharging grid operator (ERGO) for creating a market to coordinate production and consumption of renewable energy. They argued that the ERGO model could overcome both problems of GHG emissions and power fluctuations (CO_2 emissions produced by the private transport sector and fluctuations in supply made by intermittent resources) by converting EVs to distributed storage devices for electricity. An introduction of V2G distributed power sources, IT intelligence to the grid, creating virtual power plants through distributed resources, and providing new applications for carbon credits have been stated as associated benefits that are achieved by the ERGO model.

Weiller (2011) applied a model to examine the impacts of different charging scenarios for PHEVs in the USA on electricity demand in view of time of day and charging place (owner's home, workplace, and public area). The results show that the possibility of charging in places other than home increases the fraction of daily energy use of PHEVs from 24 to 29 % (1.5–2.0 kWh/day). Based on the results, PHEV-20 (vehicles with a 20-mile range) makes a shift of 45–65 % of miles traveled to electricity compared to 65–80 % for PHEV-40. Furthermore, PHEVs enable US drivers to cut gasoline consumption by more than 50 % by shifting 45–77 % of miles traveled to electricity. Weiller indicated that PHEVs could be considered as a cost-effective solution when we compare electricity costs of about \$0.03/mile (\$0.13/kWh) with gasoline costs of \$0.12/mile (\$3/gallon). Environmental and transportation policies in terms of supporting EVs play a key role in measuring their effects on electricity networks in the future. Financial incentives and governmental policy regarding carbon tax can influence early and comprehensive implementation of EVs strongly.

3.3.2 Combined Heat and Power

Cogeneration or combined heat and power is the use of heat and electric power together. This is expected to enable substantial gains in efficiency. Most power distribution companies supply only electricity, not hot water or steam. Considering that almost 30–40 % of a country's total energy load is used for heating, the lack of possibility to purchase thermal energy is not fortune. CHP is an efficient use of fuel when some energy is discarded as waste heat. It captures some or all waste energy as a byproduct for heating. In Reykjavik (capital city of Iceland) and New York, end users are able to purchase both electricity and thermal energy from a utility company (Tester et al. 2005). A good example of cogeneration is the CHP unit in Avedore, Denmark. It was built as a multifuel plant with the possibility of using coal, natural gas, and biomass. But the Danish government prohibited burning coal in 1996, and Avedore switched to gas and biomass, achieving an efficiency of 55 % in CHP (Ngô and Natowitz 2009). Shipley et al. (2008) calculated that CHP capacity of the USA would increase to 20 % by 2030, which would lead to 5.3 quads reduction in energy consumption and 848 MMT carbon dioxide emissions. Based on his findings, the USA saved more than 1.9 quadrillion British thermal units (quads) of fuel consumption and 248 million metric tons carbon dioxide emissions by employing CHP. According to *WEO 2012*, the average efficiency of power plants is 41 % worldwide, and almost 60 % of primary energy is converted to waste heat (IEA 2012b). CHP could transform a significant part of waste heat to economic value for industrial processes or heating in residential and commercial buildings. It is expected that new CHP units could improve energy efficiency to more than 85 %.

Maidment and Tozer (2002) investigated the application of combined cooling heat and power (CCHP) for supermarkets in the UK and compared it to the energy saving/capital costs of conventional technology. The results show that CCHP is able to provide significant primary energy and reduce CO_2 emissions compared to conventional schemes, but it should compete with more efficient technologies in the long term. Maidment and Tozer (2002) argued that new technologies such as fuel cells could provide more improvement in energy efficiency for CCHP in refrigeration in the long term.

Hawkes and Leach (2007) examined cost-affecting operating strategies of three alternative micro-CHP technologies—Stirling engine, gas engine, and solid oxide fuel cell–based (SOFC) system—for residential application in the UK. They evaluated the economic and environmental attributes of three operating strategies—heat-led, electricity-led, and least-cost—when applied to the abovementioned technologies. The results show that SOFC-based systems have the maximum operating costs and CO_2 emission reduction, followed by the least-cost operating strategy. You et al. (2009c) examined the electricity export capability of aggregated micro-CHP units as a virtual power plant (VPP) through participation in the electricity wholesale market without any difference compared to conventional power plants. They found that export capability of micro-CHP systems strongly depends on technical parameters, associated energy price, and demand profile. Based on an applied model, they found that the marginal price for a micro-CHP system is higher than the spot price for most part of a year. Furthermore, they argued that the variable price for electricity export could be better than the fixed price.

Kiviluoma and Meibom (2010) applied a model to analyze the impact of variable power generation by wind turbines and utilizing EVs to store electricity for later consumption in order to enhance flexibility of the power grid. Based on the results, CHP units could make a good potential for power systems to be flexible in terms of production and use of heat. Christidis et al. (2012) investigated the contribution of heat storage to optimize CHP units in liberalized electricity markets, applying a model to measure the economic potential and optimal capacity of heat accumulators. They concluded that separating electricity production and heat demand could provide profitable payback periods for storage devices in the proposed energy system.

3.3.3 Virtual Power Plants

Virtual power plant (VPP) is a cluster of distributed energy resources such as micro-CHP, wind turbines, and solar photovoltaic panels, which are controlled and managed by a central control unit. The term *distributed energy resources (DER)* can be used for fossil or renewable energy fuels. The DER system was conceived to overcome the energy waste problem due to long-distance and transmission losses. Therefore, DERs are generally located close to distribution networks. The concept of VPP is used for DER integration. According to Europe FENIX project, "A Virtual Power Plant (VPP) aggregates the capacity of many diverse DERs, it creates a single operating profile from a composite of the parameters characterizing each DER and can incorporate the impact of network on aggregate DERs output. A VPP is a flexible representation of a portfolio of DERs that can be used to make contracts in the wholesale market and to offer services to the system operator. There are two types of VPP: the Commercial VPP (CVPP) and the Technical VPP (TVPP). DERs can simultaneously be part of both a CVPP and a TVPP" (Kieny et al. 2009). Commercial VPP is defined as a portfolio that could be used by DER to participate in electricity markets. CVPPs can represent DER from any geographic location in the electricity network. Technical VPP enables operators to facilitate deploying DER energy capacity and optimize power balance in the system with the minimum cost (Pudjianto et al. 2007).

The share of distributed generation (DG) in electricity networks is increasingly gaining importance, and VPP is considered as an emerging technology to enhance energy efficiency. Schulz et al. (2005) analyzed the technical and economic feasibility of operating a VPP with micro-CHP units. They explained that due to Germany's plan to abandon nuclear power plants until 2020 and building new plants with a capacity of 40 GW, a part of this new capacity should be renewable energy sources and CHP utility, which are considered as distributed generation units. VPP is an alternative to manage these units as lack of control is a big advantage for renewable energy technology. Based on their findings, the power generated by an individual owner is too small to supply because the size of power output should be 30 MW or higher based on the existing regulation. VPP operators can integrate a large number of DERs and provide 30 MW by aggregating 6,000 micro-CHP units of 5 kW power output. Schulz et al. (2005) estimated that every unit is charged 300 Euros for connection to an integrated system and return flows are divided as a share of 45 % for the unit's owner and 55 % for the operator. Ruiz et al. (2009) applied a model to manage a VPP constituted of a large number of customers with controlled home appliances in order to optimize load reduction over a certain time schedule. They estimated the size of load reduction provided by each unit of DGs in the energy market and at the same time helping optimize network congestion and balance between supply and demand. The capability of applying an optimization algorithm on the actual system has been tested in northern Spain, and results show that contribution of a large number of customers with controlled appliances could improve energy efficiency effectively.

Jansen et al. (2010) examined an architecture and communication pattern for employing a large number of electric vehicles to be integrated in a VPP system. They argued that EVs have a good potential to be a part of electricity networks if a fleet of vehicles are managed appropriately and proposed integrated VPP constituted electric vehicles. Intelligence is required to optimize the charging of EV batteries in order to manage integration of EVs into the electricity network. You et al. (2009a, b) proposed a market-based VPP model constituted of DER units that have access to electricity markets. Based on the model, general bidding and price signals are considered as two operation scenarios performed by one market-based VPP. Wille-Haussmann et al. (2010) developed a mathematical optimization model for the management of CHPs. Considering the main task of a VPP to increase generated electricity by DG units, operation of individual generators should be optimized ,and then their contribution to the output of VPP is calculated. They applied this model for a local heating system constituted of 5 CHP units, and the results show a 10 % increase in the benefits compared to a general CHP system. An extended version of this model could be used in the European Energy Exchange to optimize the management of CHP in the form of VPP.

3.3.4 Smart Meter

The most important objective for power generation companies in demand-side management is to reduce peak demand during a certain period of hours. In this regard, smart meter is a device to record consumption of electricity in hourly intervals, and the information is monitored by utility and customers. Smart meter is able to have two-way communication and intelligence management for home appliances. Smart meters allow for demand management policies to reduce the consumption and generation of electricity (Heshmati 2014). Hartway et al. (1999) examined the application of smart meters and customer choice control in order to show that the time-of-use (TOU) strategy can be beneficial for a utility company. The results show that the TOU rate option could make a 107 kWh energy saving for each customer per year. They calculated an annual bill saving of $77 for customers and cost saving of $134 per customer for the utility company. Applying smart meters could enable electricity networks to make a significant change in energy efficiency.

Faruqui et al. (2007) calculated a 5 % decrease in the US peak demand through installation of advanced metering infrastructure, which can make a substantial saving in generation, transmission, and distribution costs—enough to eliminate 625 peak load power plants and associated infrastructure, roughly $3 billion in a year. Karnouskos et al. (2007) indicated that "smart meters empowered and advanced metering infrastructure which is able to react almost in real time, provide fine-grained energy production or consumption info and adapt its behavior proactively." They argued that smart meters could provide new opportunities in electricity networks and system integration through processing data and making decisions based on capabilities. This role enables managers and policymakers to take advantage of real-time data. It is forecast that smart meters could be the gateway of home appliance communication through the Internet and will enable the use of advanced communication capabilities in the future. They concluded that a combination of energy domain and ICT could provide a great opportunity for business in coming years.

Faruqui et al. (2010) quantified long-term cost–benefit of investing in dynamic pricing and installing smart meters for the EU. They estimated that installation cost of smart meters would be 51 billion Euros compared to operational saving of 26–41 billion, which would make a gap of around 10–25 billion Euros. They argued that smart meters have the capability to cover this gap by dynamic pricing and reducing peak demand. They suggested that policymakers and utility companies are able to increase adoption rates by applying innovative policies in order to encourage customers for more participation. It is expected that the amount of saving due to reduction in associated capacity and transmission costs will be 67 billion if 80 % of customers reduce their electricity consumption during peak hours.

Depuru et al. (2011) examined different features and technologies that can be integrated with smart metering in order to figure out requirements to implement a network appropriate for smart grid communication. The worldwide deployment of smart meters is estimated to be almost 212 million units by 2014. The study indicated that home area network (HAN) technologies could support PHEVs and DG units in communication networks. Considering a significant growth rate of PHEV penetration in the future, there could be a great demand for smart meter application. Due to increasing fuel prices and high initial costs of developing conventional infrastructure for the supply side of electricity networks, energy efficiency and implementing demand response (DR) programs through smart metering is an attractive option for policymakers and utility company managers. Baltimore Gas and Electric Company has estimated the capital cost of a DR program at $165/kW, which is much less than building new peak demand generation facilities at $600–800/kW (Vojdani 2008).

Krishnamurti et al. (2012) discussed consumers' expectations and their behavioral decisions, applying a model to measure the impact of smart meter installation on beliefs about smart metering. According to the results, there is a misconception among consumers about the impact of smart metering deployment. The study suggests that this misconception could be eased by electricity utilities by explaining the potential risks and benefits clearly and taking action to communicate safeguards regarding concerns about privacy and loss of control. McKenna et al. (2012) analyzed consumer privacy concerns about smart metering and some applications of smart meters' data required for electricity industries. They examined how much sensitivity is acceptable for obtaining data and investigated whether deploying personal data can be minimized or avoided. Based on their results, it is suggested that applying appropriate privacy techniques could reduce power supply requirements for sensitive smart metering. This is crucial because privacy concerns have a strong potential to delay smart meter penetration if they are not applied appropriately.

McHenry (2013) discussed technical and governance considerations for smart meter infrastructures, including technical and nontechnical requirements, cost–benefit smart meter infrastructures, and impact of smart meter installation on stakeholders (market operators, distributors, customers, and so on). He argued that the full benefits of advanced metering infrastructure (AMI) alongside other technologies enable stakeholders to take advantage of intelligent management in order to minimize cost, improve efficiency, and enable remote monitoring. Although the potential benefits of AMI could be significant, it is stated that the scale of smart meter investment is considered as an unprecedented challenge for policymakers because of certain uncertainties of associated benefits such as retailer investment on smart metering, transferring investment cost to customers, and residential behavior influenced by current tariff and payment methods.

References

Andersen PH, Mathews JA, Rask M (2009) Integrating private transport into renewable energy policy: the strategy of creating intelligent recharging grids for electric vehicles. Energy Policy 37(7):2481–2486

Benitez LE, Benitez PC, Van Kooten GC (2008) The economics of wind power with energy storage. Energy Economics 30(4):1973–1989

Blanco MI (2009) The economics of wind energy. Renew Sust Energ Rev 13(6):1372–1382

Bodansky D (2005) Costs of electricity. In: Nuclear energy: principles, practices, and prospects. Springer, New York, pp 559–577

BP (2012) BP Statistical Review of World Energy. www.bp.com/statisticalreview/

Branker K, Pathak M, Pearce J (2011) A review of solar photovoltaic levelized cost of electricity. Renew Sust Energ Rev 15(9):4470–4482

Chamorro CR, Mondéjar ME, Ramos R, Segovia JJ, Martín MC, Villamañán MA (2012) World geothermal power production status: energy, environmental and economic study of high enthalpy technologies. Energy 42(1):10–18

Christidis A, Koch C, Pottel L, Tsatsaronis G (2012) The contribution of heat storage to the profitable operation of combined heat and power plants in liberalized electricity markets. Energy 41(1):75–82

Connolly D, Lund H, Finn P, Mathiesen BV, Leahy M (2011) Practical operation strategies for pumped hydroelectric energy storage (PHES) utilising electricity price arbitrage. Energy Policy 39(7):4189–4196

Crawford R (2009) Life cycle energy and greenhouse emissions analysis of wind turbines and the effect of size on energy yield. Renew Sust Energ Rev 13(9):2653–2660

Deane JP, O'Gallachóir B, McKeogh E (2010) Techno-economic review of existing and new pumped hydro energy storage plant. Renew Sust Energ Rev 14(4):1293–1302

Depuru SSSR, Wang L, Devabhaktuni V (2011) Smart meters for power grid: challenges, issues, advantages and status. Renew Sust Energ Rev 15(6):2736–2742

Devine Jr, W (1977) Energy analysis of a wind energy conversion system for fuel displacement. Institute for Energy Analysis, Oak Ridge

Ehnberg S, Bollen MH (2005) Reliability of a small power system using solar power and hydro. Electr Power Syst Res 74(1):119–127

Faruqui A, Harris D, Hledik R (2010) Unlocking the € 53 billion savings from smart meters in the EU: how increasing the adoption of dynamic tariffs could make or break the EU's smart grid investment. Energy Policy 38(10):6222–6231

Faruqui A, Hledik R, Newell S, Pfeifenberger H (2007) The power of 5 percent. Electr J 20(8):68–77

Feltrin A, Freundlich A (2008) Material considerations for terawatt level deployment of photovoltaics. Renew Energy 33(2):180–185

Ford A (1995) The impacts of large scale use of electric vehicles in Southern California. Energy Build 22(3):207–218

Frankl P, Masini A, Gamberale M, Toccaceli D (1997) Simplified life-cycle analysis of PV systems in buildings: present situation and future trends. INSEAD, Centre for the Management of Environmental Resources

Frick S, Kaltschmitt M, Schröder G (2010) Life cycle assessment of geothermal binary power plants using enhanced low-temperature reservoirs. Energy 35(5):2281–2294

Fridleifsson IB, Freeston DH (1994) Geothermal energy research and development. Geothermics 23(2):175–214

Fthenakis VM, Kim HC (2007) Greenhouse-gas emissions from solar electric-and nuclear power: a life-cycle study. Energy Policy 35(4):2549–2557

Fthenakis VM, Kim HC, Alsema E (2008) Emissions from photovoltaic life cycles. Environ Sci Technol 42(6):2168–2174

Fthenakis V, Mason JE, Zweibel K (2009) The technical, geographical, and economic feasibility for solar energy to supply the energy needs of the US. Energy Policy 37(2):387–399

Gagnon L, van de Vate JF (1997) Greenhouse gas emissions from hydropower: the state of research in 1996. Energy Policy 25(1):7–13

Gevorkian P (2012) Large scale solar power systems: construction and economics. Cambridge University Press, Cambridge

Gipe P (1995) Wind energy comes of age, vol 4. Wiley, New York

Gordon J (1987) Optimal sizing of stand-alone photovoltaic solar power systems. Solar Cells 20(4):295–313

Haack BN (1981) Net energy analysis of small wind energy conversion systems. Appl Energy 9(3):193–200

Hartway R, Price S, Woo C (1999) Smart meter, customer choice and profitable time-of-use rate option. Energy 24(10):895–903

Hawkes A, Leach M (2007) Cost-effective operating strategy for residential micro-combined heat and power. Energy 32(5):711–723

Heshmati A (2014) An empirical survey of the ramification of a green economy. IZA Discus Pap 2014:8078

Huo M, Zhang X, He J (2011) Causality relationship between the photovoltaic market and its manufacturing in China, Germany, the US, and Japan. Front Energy 5(1):43–48

IEA (2012a) Energy technology perspectives 2012. OECD Publishing, Paris

IEA (2012b) World energy outlook 2012. OECD Publishing, Paris

IEA (2012c) Energy policies of IEA countries: Denmark 2011. OECD Publishing, Paris

IEA (2012d) Medium-term renewable energy market report 2012. OECD Publishing, Paris

Ito M, Kato K, Komoto K, Kichimi T, Kurokawa K (2008) A comparative study on cost and life-cycle analysis for 100 MW very large-scale PV (VLS-PV) systems in deserts using m-Si, a-Si, CdTe, and CIS modules. Prog Photovolt Res Appl 16(1):17–30

Jacobsson S, Bergek A (2004) Transforming the energy sector: the evolution of technological systems in renewable energy technology. Ind Corp Chang 13(5):815–849

Jäger-Waldau A (2006) European Photovoltaics in world wide comparison. J Non-Cryst Solids 352(9):1922–1927

Jansen B, Binding C, Sundstrom O, Gantenbein D (2010) Architecture and communication of an electric vehicle virtual power plant. Paper presented at the Smart Grid Communications (SmartGridComm), 2010 First IEEE International Conference on, IBM Research-Zurich, Switzerland

Kaldellis J, Kapsali M, Kavadias K (2010) Energy balance analysis of wind-based pumped hydro storage systems in remote island electrical networks. Appl Energy 87(8):2427–2437

Kapsali M, Kaldellis J (2010) Combining hydro and variable wind power generation by means of pumped-storage under economically viable terms. Appl Energy 87(11):3475–3485

Karnouskos S, Terzidis O, Karnouskos P (2007) An advanced metering infrastructure for future energy networks. In: New technologies, mobility and security. Springer, New York, pp 597–606

Kaya E, Zarrouk SJ, O'Sullivan MJ (2011) Reinjection in geothermal fields: a review of worldwide experience. Renew Sust Energ Rev 15(1):47–68

Kempton W, Letendre SE (1997) Electric vehicles as a new power source for electric utilities. Transp Res Part D Transp Environ 2(3):157–175

Kempton W, Tomić J (2005) Vehicle-to-grid power implementation: from stabilizing the grid to supporting large-scale renewable energy. J Power Sources 144(1):280–294

Kieny C, Berseneff B, Hadjsaid N, Besanger Y, Maire J (2009) On the concept and the interest of Virtual Power plant: some results from the European project FENIX. Paper presented at the Power & Energy Society General Meeting, 2009. PES'09. IEEE

Kiviluoma J, Meibom P (2010) Influence of wind power, plug-in electric vehicles, and heat storages on power system investments. Energy 35(3):1244–1255

Klaassen G, Miketa A, Larsen K, Sundqvist T (2005) The impact of R&D on innovation for wind energy in Denmark, Germany and the United Kingdom. Ecol Econ 54(2):227–240

Kolhe M, Kolhe S, Joshi J (2002) Economic viability of stand-alone solar photovoltaic system in comparison with diesel-powered system for India. Energy Economics 24(2):155–165

Korpaas M, Holen AT, Hildrum R (2003) Operation and sizing of energy storage for wind power plants in a market system. Int J Electr Power Energy Syst 25(8):599–606

Krishnamurti T, Schwartz D, Davis A, Fischhoff B, de Bruin WB, Lave L, Wang J (2012) Preparing for smart grid technologies: a behavioral decision research approach to understanding consumer expectations about smart meters. Energy Policy 41:790–797

Kubiszewski I, Cleveland CJ, Endres PK (2010) Meta-analysis of net energy return for wind power systems. Renew Energy 35(1):218–225

Lehner B, Czisch G, Vassolo S (2005) The impact of global change on the hydropower potential of Europe: a model-based analysis. Energy Policy 33(7):839–855

Lenzen M, Munksgaard J (2002) Energy and CO2 life-cycle analyses of wind turbines—review and applications. Renew Energy 26(3):339–362

Lenzen M, Wachsmann U (2004) Wind turbines in Brazil and Germany: an example of geographical variability in life-cycle assessment. Appl Energy 77(2):119–130

Liberman EJ (2003) A life cycle assessment and economic analysis of wind turbines using Monte Carlo simulation. DTIC Document

Lin GT (2011) The promotion and development of solar photovoltaic industry: discussion of its key factors. Distrib Generat Altern Energy J 26(4):57–80

Lund H, Kempton W (2008) Integration of renewable energy into the transport and electricity sectors through V2G. Energy Policy 36(9):3578–3587

Lund JW, Freeston DH, Boyd TL (2005) Direct application of geothermal energy: 2005 worldwide review. Geothermics 34(6):691–727

Maidment G, Tozer R (2002) Combined cooling heat and power in supermarkets. Appl Therm Eng 22(6):653–665

Martinot E, Sawin J (2012) Renewables global status report. Renewables 2012 Global Status Report, REN21. http://www.martinot.info/REN21_GSR2012.pdf

McHenry MP (2013) Technical and governance considerations for advanced metering infrastructure/smart meters: technology, security, uncertainty, costs, benefits, and risks. Energy Policy 59:834–842

McKenna E, Richardson I, Thomson M (2012) Smart meter data: balancing consumer privacy concerns with legitimate applications. Energy Policy 41:807–814

Monteiro C, Ramirez-Rosado IJ, Fernandez-Jimenez LA (2013) Short-term forecasting model for electric power production of small-hydro power plants. Renew Energy 50:387–394

Murphy H, Niitsuma H (1999) Strategies for compensating for higher costs of geothermal electricity with environmental benefits. Geothermics 28(6):693–711

Nawaz I, Tiwari G (2006) Embodied energy analysis of photovoltaic (PV) system based on macro- and micro-level. Energy Policy 34(17):3144–3152

Ngô C, Natowitz JB (2009) Our energy future: resources, alternatives and the environment, vol 11. Wiley, Hoboken

Nieuwenhout FDJ, Van Dijk A, Lasschuit PE, Van Roekel G, Van Dijk VAP, Hirsch D, Arriaza H, Sharma BD, Wade H (2001) Experience with solar home systems in developing countries: a review. Prog Photovolt Res Appl 9(6):455–474

OECD (2010) Projected costs of generating electricity 2010. OECD Publishing, Paris

Oliver M, Jackson T (1999) The market for solar photovoltaics. Energy Policy 27(7):371–385

Paish O (2002) Small hydro power: technology and current status. Renew Sust Energ Rev 6(6):537–556

Pudjianto D, Ramsay C, Strbac G (2007) Virtual power plant and system integration of distributed energy resources. Renew Power Gener IET 1(1):10–16

Purkus A, Barth V (2011) Geothermal power production in future electricity markets—a scenario analysis for Germany. Energy Policy 39(1):349–357

Raadal HL, Gagnon L, Modahl IS, Hanssen OJ (2011) Life cycle greenhouse gas (GHG) emissions from the generation of wind and hydro power. Renew Sust Energ Rev 15(7):3417–3422

Raugei M, Frankl P (2009) Life cycle impacts and costs of photovoltaic systems: current state of the art and future outlooks. Energy 34(3):392–399

Ruiz N, Cobelo I, Oyarzabal J (2009) A direct load control model for virtual power plant management. Power Syst IEEE Trans 24(2):959–966

Saner D, Juraske R, Kübert M, Blum P, Hellweg S, Bayer P (2010) Is it only CO2 that matters? A life cycle perspective on shallow geothermal systems. Renew Sust Energ Rev 14(7):1798–1813

Sarver T, Al-Qaraghuli A, Kazmerski LL (2013) A comprehensive review of the impact of dust on the use of solar energy: history, investigations, results, literature, and mitigation approaches. Renew Sust Energ Rev 22:698–733

Schleisner L (2000) Life cycle assessment of a wind farm and related externalities. Renew Energy 20(3):279–288

Schulz C, Roder G, Kurrat M (2005) Virtual Power Plants with combined heat and power micro-units. Paper presented at the Future Power Systems, 2005 International Conference on, Technical University Braunschweig, Germany

Shipley MA, Hampson A, Hedman MB, Garland PW, Bautista P (2008) Combined heat and power: effective energy solutions for a sustainable future. Oak Ridge National Laboratory (ORNL), Oak Ridge

Shum KL, Watanabe C (2007) Photovoltaic deployment strategy in Japan and the USA—an institutional appraisal. Energy Policy 35(2):1186–1195

Sinha A (1993) Modelling the economics of combined wind/hydro/diesel power systems. Energy Convers Manag 34(7):577–585

Steenhof PA, McInnis BC (2008) A comparison of alternative technologies to de-carbonize Canada's passenger transportation sector. Technol Forecast Soc Chang 75(8):1260–1278

Stefansson V (2002) Investment cost for geothermal power plants. Geothermics 31(2):263–272

Sundararagavan S, Baker E (2012) Evaluating energy storage technologies for wind power integration. Sol Energy 86(9):2707–2717

Tester JW, Drake EM, Driscoll MJ, Golay MW, Peters WA (2005) Sustainable energy: choosing among options. The MIT Press, Cambridge, MA

Tomić J, Kempton W (2007) Using fleets of electric-drive vehicles for grid support. J Power Sources 168(2):459–468

Tremeac B, Meunier F (2009) Life cycle analysis of 4.5 MW and 250 W wind turbines. Renew Sustain Energy Rev 13(8):2104–2110

Tsoutsos T, Frantzeskaki N, Gekas V (2005) Environmental impacts from the solar energy technologies. Energy Policy 33(3):289–296

Vojdani A (2008) Smart integration. Power Energ Mag IEEE 6(6):71–79

Wagner HJ, Pick E (2004) Energy yield ratio and cumulative energy demand for wind energy converters. Energy 29(12):2289–2295

Weiller C (2011) Plug-in hybrid electric vehicle impacts on hourly electricity demand in the United States. Energy Policy 39(6):3766–3778

Wille-Haussmann B, Erge T, Wittwer C (2010) Decentralised optimisation of cogeneration in virtual power plants. Sol Energy 84(4):604–611

Wirl F (1989) Optimal capacity expansion of hydro power plants. Energy Economics 11(2):133–136

Yang CJ, Jackson RB (2011) Opportunities and barriers to pumped-hydro energy storage in the United States. Renew Sust Energ Rev 15(1):839–844

You S, Træholt C, Poulsen B (2009a) Generic virtual power plants: management of distributed energy resources under liberalized electricity market. Paper presented at the Advances in Power System Control, Operation and Management (APSCOM 2009), 8th International Conference on, Technical University of Denmark, Kgs. Lyngby

You S, Træholt C, Poulsen B (2009b) A market-based virtual power plant. Paper presented at the Clean Electrical Power, 2009 International Conference on, Technical University of Denmark, Kgs. Lyngby

You S, Traholt C, Poulsen B (2009c) A study on electricity export capability of the μCHP system with spot price. Paper presented at the Power & Energy Society General Meeting, 2009. PES'09. IEEE

Chapter 4
Regulatory Frameworks for Renewable Energy Sources

4.1 Introduction

Regulation refers to the process of making, monitoring, and enforcing rules that are established by a state. These rules are mandated by a state to produce appropriate and desirable outcomes. According to Gunningham et al. (1998), regulatory instruments include command regulation, control regulation, market creation, education instruments, information instruments, economic instruments, self-regulation, voluntarism, property rights, fiscal instruments and charge systems, financial instruments, liability instruments, performance bonds, deposit refund systems, and the removal of perverse incentives.

The interaction between players and organizations (either state or nonstate) in markets and their reciprocal influence should be considered in effective policymaking. Environmental regulations could affect the commercial policies and strategic decisions made by companies. According to Bosselmann and Richardson (1999), regulations provide "planning frameworks for resource use and environmental protection, limitation on market entry and exit, specifications relating to the methods of production, and controls on the quality of the products supplied." Interactions take place between government policies, commercial policies, and environmental policies. For example, in the case of liberalization, some governments may reduce the restrictions imposed by environmental regulations in order to help domestic companies compete with their international rivals.

In this chapter, the main objective is to conduct a comprehensive review of the literature on regulation frameworks that are employed around the world and are related to the use of renewable energy sources. An attempt is made to classify the frameworks and identify the strengths and weaknesses of different regulatory frameworks. Finally, based on the findings, a set of optimal frameworks is proposed, and procedures to facilitate their effective implementation are suggested.

In our discussion, the regulatory frameworks are related to different dimensions of renewable energy sources. These dimensions include environmental policy,

© Springer Science+Business Media Singapore 2015
A. Heshmati et al., *The Development of Renewable Energy Sources and its Significance for the Environment*, DOI 10.1007/978-981-287-462-7_4

public policy, commercial policy, and economic policy. These regulations are required to overcome market failures and produce desirable outcomes. Such regulatory interventions could be classified into three main groups: economic regulation, regulation of anticompetitive behavior, and social regulation (Bhattacharyya 2011). A broader classification, which we follow in this chapter, includes support policy, market regulation, technology transfer, barriers, and international regulatory policies. It is important to mention that regional and international regulatory frameworks influence national environment regulation. Examples of this case are the different scenarios defined by the IEA for OECD countries and by the Kyoto Protocol.

4.2   Support Policy

Although the deployment of renewable energy has increased during recent decades, fossil fuels supported by subsidies remain dominant as primary global energy sources. Support policies have also played a key role in enhancing renewable energy consumption. Therefore, these policies could direct a society's use of energy for power generation, transportation, heating, and so on. For example, regulatory reforms in China are expected to push energy consumption upward from approximately 130 bcm in 2011 to 545 bcm in 2035 (IEA 2012b). Regarding renewable energies in addition to parameters such as decreasing technology costs caused by advancement and economies of scale, the rapid growth of renewable energy has been driven mainly by supporting policies. Because some sources of energy such as natural gas and coal are available in the market at lower prices, renewable energy could not be economical without government support. There may be different drivers of support policies based on government priorities including energy security, economic effects, and carbon dioxide reduction, which have been discussed in Chap. 2. Support policies could be applied from the research stage to commercialization for both the supply side (i.e., academia, research centers, and firms) and the demand side (i.e., consumers, public and private sectors, imports, and exports) (IEA 2012a).

Many OECD countries have implemented national strategies to support sustainable development through environmentally friendly technological advances. These strategies deal with different objectives and cover a wide range of policies including the environment, science and technology, transport, competition, and energy (OECD 2011). As discussed in Chap. 3, different kinds of technology are used to overcome difficulties caused by climate change. Concern about the effects of climate change and the depletion of fossil fuel reserves has urged many governments to design and implement policies to support the spillover of renewable energy technologies to different areas of use. By early 2012, at least 109 countries had applied some kind of renewable power support policies, which was an increase from the 96 countries reported in the *Renewables Global Status Report* in 2011 (Martinot and Sawin 2012). These policies are performed at the state or national level. Table 4.1 shows renewable energy support policies in the EU-15 countries.

Table 4.1 Renewable energy support mechanism in EU-15 countries

	Regulatory policies						Fiscal incentives				Public finance	
	Feed-in tariff	Quota obligation	Net metering	Biofuels obligation	Heat obligation	Tradable REC	Capital subsidy	Investment or production tax credit	Other taxes	Energy production payment	Public investment	Public competitive bidding
Austria	n			n		n	n	n			n	
Belgium		s	n	n		n	s	n	n			n
Denmark	n		n	n		n	n	n	n		n	n
Finland	n			n		n	n		n	n		
France	n			n	n		n	n	n		n	n
Germany	n			n	n		n	n	n		n	
Greece	n			n			n	n	n		n	
Ireland	n			n	s	n						n
Italy	n	n	n	n	n	n						n
Luxembourg	n						n					
Netherlands	n			n		n	n	n	n	n	n	
Portugal	n	n	n	n	n		n	n	n		n	n
Spain	n			n	n		n	n	n		n	
Sweden		n		n		n	n	n	n		n	
UK	n	n		n	n	n	n	n	n	n	n	

Source: Renewable Global Status Report, 2012
Notes: n national-level policy, *s* state/provincial-level policy

Governments can utilize a number of options to enhance renewable energy deployment. Sawin (2004) gave three options of instruments in the support policy mechanism: (i) supporting voluntarism through education and information; (ii) supporting environmental standards and energy taxes; and (iii) supporting renewable energy technologies directly. Education is a crucial component in using renewable energy technologies. Sawin has stated that the effectiveness of government policies depends on the design and enforcement of policies. There is no guarantee that a particular policy will be successful. Based on this finding, support policies should focus on the end to promote renewable energy technologies on a small-scale and distributed basis. Gunningham et al. (1998) divided education and information instruments into five major categories: education and training, corporate environmental reporting, community right-to-know and pollution inventories, product certification, and award schemes. Education and training are considered a crucial part of the mechanism to develop renewable energy technologies in the industry and residential sectors. They are essential in changing mindsets and facilitating customer acceptance.

Corporate environmental reporting is considered a useful practice (i.e., prepared as part of an annual report or a separate report) in enhancing the environmental protection activities performed by firms. Community right-to-know (CRTK) is used as a policy to inform communities about the environmental impacts of pollution caused by a firm's activities. Providing customers who care about energy consumption with information is facilitated by product certification. Evidence shows that ecolabeling as a form of sustainability measurement enables customers to consider environmental concerns in combination with education and information strategies as a mixed policy of a regulatory regime.

Dinica (2006) discussed support systems for the diffusion of renewable energy technology from the point of view of investors. She analyzed the investor's perspective in order to examine the potential of support policies for the spillover effects of renewable energy and found that the investor's decision-making is influenced by the risk/profitability of policies, not the kind of instrument used. The study emphasized that including the investor's perspective in examining government policies enables support to the diffusion of renewable energy technologies for the generation of electricity generation. Moreover, such policies could be more effective if they were based on risk and profitability for investors. Dinica's (2006) main argument is the importance of the investor's perspective in achieving successful outcomes. Many policymakers favor a support policy regardless of the result and relationship between the policy and the policy takers. Fouquet and Johansson (2008) examined European renewable energy policy and focused on support mechanisms for electricity. They argued that renewable energy could not be developed in energy markets and that some support mechanism should be considered for expanding the use of renewable energy to enhance energy security, reduce greenhouse gases, and improve economies. The results showed that the feed-in tariff policy is more effective than tradable green certificates are.

Verbruggen (2009) evaluated the performance of a support system for renewable energy sources to generate electricity. He compared the results of the Flemish system with simulations applied in Germany. The findings showed that the tradable certificate system in Flanders had not been developed properly. The market does not have the required functions such as economic structure, professional constructors, and supervisors. Verbruggen also indicated three crucial objectives that should be clarified before designing a support policy: target setting, qualification of RES-E (electricity produced from renewable energy) sources and technologies, and the robustness of achieved levels of effectiveness. He stated that the qualification of RES-E sources and technologies is the basic task in establishing a functional market.

In a recent research, Dinica (2011) discussed how production cost is influenced by national–contextual factors, observing that a framework is required to help policymakers in making decisions regarding renewable energy support. The study emphasized that "policy decisions regarding the long-term strategy for governmental RET (renewable energy technology) price support should not be exclusively based on experience curve." Dinica suggested an analytical framework for policymakers to use as a cost-effective instrument to support renewable energy technology and maximize the share of low carbon electricity in energy consumption.

del Río and Bleda (2012) compared the innovative effects of policy instruments that support renewable energy technology spillover. They found that the effects vary according to different renewable electricity instruments. However, the feed-in-tariff (FIT) policy is better than other policy instruments such as tradable green certificates. del Río and Bleda argued that market creation should be focused instead of increasing competition between different kinds of technologies because it is able to generate broad support for renewable energy technologies. Therefore, del Río and Bleda supported the idea of market formation and believed that market creation could indirectly influence other functions.

Steinbach et al. (2013) investigated policies that were formed to develop renewable heating and cooling and applied a quantitative analysis to evaluate the costs and benefits of different scenarios in six member states (Austria, Greece, the Netherlands, Lithuania, Poland, and the UK). The results showed that the intended target could be achieved by harmonized obligations and decreased generation costs, which were facilitated by better resource allocation. Steinbach et al. (2013) have categorized the levels of policy harmonization to common target settings, central coordination, convergence of instrument type, and convergence of instrument design. Their quantitative assessment was based on the following criteria: cost-optimal resource allocation, enforced target compliance, minimization of transaction cost, minimization of total policy cost, and avoidance of market distortion in order to support the idea of a harmonized European internal market.

4.3 Market Regulation

The regulatory mechanism has a central role in developing the generation of power using renewable energy sources. Governmental support is required to facilitate new technology applied to developing renewable energy. The lack of a regulatory framework and market environment to support new technology and investment increases the probability of market failure. In addition to grid modernization, efforts to enhance energy efficiency and adapt current policy, regulatory frameworks, and market environments are crucial to support investment in new technology. Therefore, it will be a major challenge for stakeholders in the electricity sector. A survey by the IEA found that technology improvement together with targeted policy and regulation are essential in achieving the various scenario targets by 2050 (IEA 2012a).

Based on evidence from the US manufacturing sector, Jaffe et al. (1995) analyzed the arguments regarding environmental regulation and the negative impact on market competitiveness caused by significant costs and slow productivity growth and found little evidence to support this argument. They stated that additional costs imposed by environmental regulations are a small portion of unit costs for most heavily regulated industries. Jaffe et al. believed that international differences between environmental regulations in the USA and other countries are not considered a threat to the market competitiveness of the US industry. Jaffe and Palmer (1997) investigated the effects of environmental regulations in a panel of manufacturing industries and found that environmental expenditures have a significantly positive effect on R&D. The results showed that the severe effects of environmental regulation stimulate innovative activity in firms. They argued that there is little evidence to support the idea that the inventive output of industries as measured by successful patent applications is related to compliance cost. Murphy and Gouldson (2000) analyzed interactions between environmental policy and industrial innovation and argued that clean technologies, organizational change, and the adoption of long-term radical innovation is the only solution to achieve targets in ecological modernization. They indicated that environmental regulation is able to both force and facilitate the adoption of innovation.

Ackermann et al. (2001) evaluated existing government instruments and market schemes that support implementation of renewable energy power generation with the aim of analyzing market regulations that stimulate distributed generation resources. They examined seven different instruments to find a solution to the issue of cost reduction. The results showed that tax reduction, investment subsidies, feed-in tariffs, and net metering could be used as interim solutions but might not be able to reduce generation costs. They argued that the bidding process is one way to achieve this target but it is not useful to apply in large-scale renewable energy projects because of the high level of transaction costs. Furthermore, a combination of instruments such as fixed quotas and green certificate trading or power exchange and green pricing may result in similar cost reductions. Gunningham et al. (1998) have emphasized this concept. Based on a policy framework and target setting, market instruments should be selected properly. These instruments could be used as complementary or sequential. It is recommended to avoid policies that affect each other negatively. For example, applying economic instruments such as tradable emission permits may have adverse effects if they are used with liability.

Menanteau et al. (2003) evaluated the efficiency of different policy instruments in the development of renewable energy sources. They applied a theoretical approach by comparing the price-based approach with the quantity-based approach. They found that the feed-in-tariff mechanism is more efficient than the bidding system because it enables countries to achieve targets regarding renewable energy development. Based on these results, an increasing number of countries may use the quota-based green certificate trading system in order to achieve their targets through a cost-effective solution. Anton et al. (2004) examined the impact of market regulation on environmental management systems (EMS) by applying standard Poisson and negative binomial models. EMS is constituted by different environmental management practices such as environmental policy, training and rewarding workers to find solutions to reduce pollution, setting internal standards, and doing internal environmental audits. The results showed that public policy is able to prevent toxic pollution by creating a regulatory framework and market-based pressures imposed by EMSs.

Wang (2006) evaluated the development of renewable energy policymaking in Sweden including the policy context and changes in policy instruments in order to analyze successes and failures in regulations. The Swedish government faced a dilemma in supporting renewable energy development and phasing out nuclear power: first, there was a political decision to replace nuclear power with renewable energy, and second, they were concerned about the negative effects of this policy on industrial competitiveness. Wang argued that lack of government commitment because of this uncertainty was an essential factor and was apparent in policies. Preferred short-term subsidies instead of a long-term support mechanism caused interruptions to development. Wang indicated that future renewable energy policies, particularly regarding high-cost technology, would strongly depend on nuclear policies. Such uncertainty as well as a significant share of major electricity companies in nuclear power would lead to instability in the regulations made for the development of renewable energy.

Costantini and Crespi (2008) estimated an empirical model to show that severe environmental regulations could be a positive signal for increasing investment in new technology by providing a source of comparative advantage. They found that countries with strong environmental regulation and higher innovation capability had greater export capacity and could be exporters of environmental technology. Costantini and Crespi used a gravity model to examine the determinants and transmission channels in exporting environmental technologies for renewable energy and energy efficiency from the EU to advanced and developing countries. They emphasized, "the stringency of environmental regulation supplemented by the strength of the National Innovation System is a crucial driver of export performance in the field of energy technologies." de Joode et al. (2009) analyzed the effect of the growing penetration of decentralized electricity generation (DG) on distribution system operators (DSOs). They considered network characteristics, technologies, and network management. They found that current market regulations should be improved in order to enable DSOs to continue facilitating the integration of DG in the network. Based on the results, they suggested implementing a regulatory framework that would affect both operating and capital costs of DG.

Recently, Zhao et al. (2011) analyzed policy frameworks to examine the influence of renewable energy regulations on the structure of power generation in China. The results showed a strong positive relationship between the diffusion of energy regulations and the growth rate of renewable energy deployment. They indicated that national laws, regulations, policies, and strategic plans are essential to promote the structure and penetration of renewable energy projects. Considering China's plan to achieve a share of 15 % of nonfossil fuel in primary energy consumption by 2020, Zhao et al. stated that the elasticity of policies to ascertain sustainable growth rates of renewable energy in power generation should be taken into account. Nykamp et al. (2012) applied data envelopment analysis (DEA) and stochastic frontier analysis (SFA) to examine the effects of incentive regulations on the investment decisions of power network operators to integrate renewable energy sources. The results showed that grid operators tended to avoid new investments. They also found that current regulations do not provide enough incentive for operators to invest in smart solutions. Therefore, changes in regulations are required to provide incentives for grid operators.

A summary of empirical research conducted for investigating the impact of environmental regulation on renewable energy development and their findings are reported in Table 4.2. The subjects of research are mainly focused on environmental regulations, policy, and incentive programs. The results point to small unit-cost effects of environmental regulations, their positive innovation effects, and identification of efficient policies, instruments, and conditions.

4.4   Technology Transfer

Advanced technology and investment in new technology are essential in promoting renewable energy consumption. The unit cost for power generation through renewable energy sources is more than for conventional resources such as fossil fuels. However, this cost could be reduced by advanced technology and economies of scale. "A technology transfer typically includes the transfer of the technology design as well as the transfer of the property rights necessary to reproduce the technology in a particular domestic context" (Lewis and Wiser 2007). In this regard, designing appropriate regulations to promote technological innovation is crucial. In particular, it is necessary for governments who set targets for emission reduction within a certain period to facilitate investment in low-carbon technologies from the demonstration to the commercial stage. There should be a proper link between actors in order to take advantage of interactive learning and new ideas. Moreover, appropriate capability is required to learn rapidly and effectively. Otherwise, it is not possible to transfer and apply new technology.

Various scholars including Carlsson and Jacobsson (1997), Smith (1997), and Johnson and Gregersen (1995) have examined system failures and found that infrastructure failure, transitional failure, hard institutional failure, network failure, and capability failure are considered system imperfections that could lead to system

Table 4.2 Empirical studies regarding the impact of environmental regulation on renewable energy development

Author	Subject	Result
Jaffe et al. (1995)	Environmental regulation and the competitiveness of US manufacturing	Additional cost imposed by environmental regulation is a small portion of unit cost for most regulated industries.
Jaffe and Palmer (1997)	Environmental regulation and innovation	Severe effects of environmental regulation stimulate innovation activity by firms.
Murphy and Gouldson (2000)	Environmental policy and industrial innovation	Clean technologies, organizational change, and adoption of radical innovation in the long term are the only solutions to achieve targets for ecological modernization.
Ackermann et al. (2001)	Government and market-driven programs	Combination of instruments such as fixed quotas and green certificate trading or power exchange and green pricing may have a cost-reduction effect.
Menanteau et al. (2003)	Choosing policies for promoting renewable energy	The feed-in-tariff mechanism is more efficient than the bidding system. Increasing number of countries may use the green certificate trading system in order to achieve their targets.
Anton et al. (2004)	Incentives for environmental self-regulation	Public policy is able to prevent toxic pollution by creating a regulatory framework and market-based pressures imposed by environmental management systems.
Wang (2006)	Analysis of policy and regulation in Sweden	Future renewable energy policies strongly depend on nuclear policies. This kind of uncertainty and the significant share of major electricity companies in nuclear power lead to instability of regulation.
Dinica (2006)	Support system for the diffusion of renewable energy technologies	An investor perspective for examining governmental policies enables support to diffusion of renewable energy technologies for electricity generation.
Costantini and Crespi (2008)	Environmental regulation and export dynamic of energy technologies	Countries with strong environmental regulation and higher innovation capability have a greater export capacity and could be exporters of environmental technology.

(continued)

Table 4.2 (continued)

Author	Subject	Result
de Joode et al. (2009)	Increasing penetration of renewable and distributed electricity generation (DG)	Current market regulation should be improved in order to enable distribution system operators to continue facilitating integration of DG in the network.
Zhao et al. (2011)	Impacts of renewable energy regulation on the power generation in China	There is a strong positive relationship between diffusion of energy regulation and the growth rate of renewable energy deployment. Elasticity of policies to ascertain sustainable growth rate of renewable energy should be taken in account.
Nykamp et al. (2012)	Standard incentive regulation	Current regulations do not provide enough incentives to operators to invest in smart solutions. A change in regulation is required to create more incentives.
del Río and Bleda (2012)	Comparing the innovation effects of support schemes	Market creation should be focused instead of increasing competition between different kinds of technologies. It could influence other functions indirectly.

failure (Klein Woolthuis et al. 2005). System imperfections are a serious problem in applying, transferring, learning, or adapting to new technological development and make it almost impossible to achieve targets in time. An analysis by the IEA showed that the process of technological change in some cases takes decades and there are limits to the rate of deploying new energy technology. In contrast, the possibility of acceleration in information technology depends on government policy (IEA 2012a).

Jaffe et al. (2005) investigated the interaction of market failures associated with environmental pollution and market failures associated with the innovation and diffusion of new technologies. Theoretical and experimental research has shown that technological advances are affected by market and regulatory incentives. Jaffe et al. believed that appropriate market regulations are able to create these incentives. Because of the lack of adequate resources to support all new technologies, the government has to focus on the commercialization of technologies that lead to increased public benefits. Technology spillover and the achievement of associated benefits could be stimulated by proper incentive instruments such as tax credits for new equipment in order to make them cost effective.

Lewis and Wiser (2007) analyzed the strategies of local industry in wind turbine manufacturing, technology acquisition, and incentives for technology transfer. They did a cross-country comparison of support policy mechanisms applied in 12 countries: Denmark, Germany, Spain, the USA, the Netherlands, the UK, Australia, Canada, Japan, India, Brazil, and China. They argued that a technology policy is influenced by short-term and long-term goals to the degree of the localization of new technology in domestic manufacturing (including assembly, components, or entire turbines). The application of appropriate policies is essential in stimulating incentives for technology localization. Lewis and Wiser indicated that a technology policy may be changed over time and incentive instruments can be adapted according to the new policy. Therefore, governments may turn to foreign direct investment for turbine manufacturing, use local transferred technology for the components, and use local manufactures for the turbines. The results showed that the annual size and stability of the market is a crucial parameter that influences policy mechanisms.

Nemet (2009) analyzed demand-pull, technology push, and government-led incentives for nonincremental technological change. This study determined the reasons for inventors of the most important inventions not being positively influenced by strong demand-pull policies. Previous research by Dosi (1988) suggested that incremental innovation responds to demand-pull policies more than technology push does. Kemp (1997) found that incentives required for incremental and nonincremental innovation vary based on their stringency. Nemet argued that there are three main reasons for the inconsistency between these policies: first, convergence on a single dominant design; second, uncertainty about the lifetime of incentives; and third, declining R&D funding. The combination of these three factors is able to offset incentive instruments.

Loock (2012) investigated a database of 249 renewable energy investment managers including banks, funds, investment advisors, private equity, and venture capitalists. He estimated three generic business models to calculate the share of preference for investors and found that investors prefer to be supported by better services than lower prices or better technology. Loock suggested that policymakers focus on policies that support service-driven business models instead of price or technology. Recently, Lema and Lema (2013) analyzed technology transfer in the clean development mechanism (CDM), which is focused on wind power. They explored how quantity flows in CDM are affected by technological capability in the host country. Based on evidence from China and India, they argued that CDM wind projects tend to take advantage of existing transfer mechanisms instead of creating new mechanisms to support low carbon technology. The results showed an important relationship between international law regarding technology transfer mechanisms and domestic technological infrastructure, which should be taken into account by policymakers.

El Fadel et al. (2013) provided knowledge management mapping in renewable energy to design a framework for defining activities in different time spans. They argued that renewable energy development in developing countries relies on financial and technological aid provided by developed countries as well as international and regional organizations. It is crucial for developing countries to facilitate appropriate capabilities (e.g., knowledge exchange, technical capability, and financial mechanism) in order to take advantage of supportive instruments.

4.5 Barriers

The most important challenge in deploying renewable energy is the intermittent character of some renewable energy sources such as wind and solar energy. This intermittency causes uncertainty in using electricity on demand. Moreover, when power generation is faced with a shortage of demand, a solution should be available. For example, Denmark is the forerunner in the generation of electricity through offshore wind turbines. It exports excess electricity to Norway and Sweden, which use mostly hydro energy. In these countries, water is stored behind dams for later use when excess electricity is transmitted by Denmark. Although hydro energy is considered a stable renewable energy source, the building of dams across rivers impacts environmental conditions (for instance, landscape). Wind turbines and economic factors such as other renewable energy sources are taken into account in environmental issues because of their impacts on flora and fauna and the noise impacts on neighborhood residents.

In terms of economics, the unit cost of power generated by renewable energy sources is generally higher compared to that of conventional sources of energy such as fossil fuels and nuclear power plants. This cost has decreased notably during the last decade because of technological advancements, which was achieved by supportive government policies. Lack of knowledge and education of consumers is another challenge in taking advantage of renewable energy sources. Social acceptance and buyer readiness are among the most important factors for market implementation. Renewable energy development policies will not be successful if they cannot influence customer acceptance at the stage of buyer readiness. All these barriers should be overcome by appropriate policies in order to develop renewable energy sources.

Painuly (2001) proposed a framework to identify the barriers for renewable energy development and find measures to bypass these barriers. In Painuly's view, the major barriers to penetrating renewable energy technology (RET) can be categorized as follows: market failure/imperfection (highly controlled energy sector, restricted access to technology, lack of information), market distortion (subsidies for fossil fuels, externalities, trade barriers), economic and financial conditions (high payback period and small size of market, capital intensive, lack of financial institutions to support RET), institutions (lack of institutions/mechanisms, lack of regulatory frameworks, clash of interests, lack of private-sector contribution), technical staff (lack of standards, lack of skilled personnel/training facilities, lack of entrepreneurs), social and cultural factors (lack of consumer acceptance of the product, lack of social acceptance of RET), and other barriers such as uncertain governmental policies and the high risk of RET. Painuly argued that barriers could be overcome through a literature survey, site visits, and interactions with stakeholders including RET industries, consumers, NGOs, experts, policymakers, and professional associations. Considering that dimensions vary across countries, the solutions to overcome the barriers may be specific to a country.

Reddy and Painuly (2004) carried out a survey among households, industries, commercial firms, and policymakers to determine their points of view regarding the barriers to the diffusion of RETs. They emphasized that "the RETs problem is a complete lack of market-based approach and it is important to initiate such a mechanism." The results showed that cost and awareness issues have significant effects, which are considered potential barriers to RET penetration. Reddy and Painuly suggested that an innovative policy network is required to overcome these barriers and that the government plays a key role in creating a competitive market based on efficiency.

Foxon et al. (2005) examined the UK innovation system for renewable energy technologies including wind, marine, solar PV, biomass, hydrogen, and micro-CHP (combined heat and power) in order to enhance the efficiency of the system. The results showed that sustainable investment is required for these technologies in order to take advantage of their potential and a stable policy framework is necessary to make it possible. Owen (2006) investigated the effects of market failure constraints on the adoption of RETs by estimating the damage costs incurred by fossil fuels and analyzing externalities of power generation in financial terms. The results showed that if external effects of fossil fuel combustion were internalized into electricity generated by conventional energy sources, a number of RETs would be competitive from the economic point of view. Owen believed that incorporating externality effects into electricity prices could stimulate the deployment of renewable energy sources for power generation. The Middle East region has great potential for generating electricity by using renewable energy sources.

Patlitzianas et al. (2006) examined RET development in the Arab states of the Persian Gulf region and found a combination of constraints including the lack of a regulatory framework, commercial skills, and required knowledge. They argued that these countries should cooperate with developed countries in order to develop renewable energy appropriately and enhance their knowledge and information through the Kyoto Protocol. They stated that a scientific and political framework should be focused and that close interaction between research centers and markets is required in order to produce new products.

Sovacool (2009a) analyzed renewable energy from the perspective of social and cultural barriers to energy efficiency technology in the USA. He argued that the clash between conventional systems of power generation and renewable energy sources is based on social concerns regarding welfare, profits, consumption, control, and trust instead of a disagreement over technology. Sovacool also believed that the US government should focus on efforts to enhance the social understanding of energy systems instead of allocating supportive incentives to promote the efficiency and technical capacity of renewable energy technologies. Sovacool (2009b) performed a survey to gather data on the common argument regarding the intermittency and unreliability of renewable power generators through 62 formal and semistructured interviews at 45 different institutions in the USA. He argued that all conventional power plants have some degree of uncertainty because of supply and demand imbalances, fuel prices, and unplanned outages. Wind and solar generators can be more efficient if they are used on large scales in geographically spaced locations such as Denmark and Spain, which have increasing shares of renewable energy contributions in continuous electricity supply. Sovacool believed that the counterargument was based on social, political, and practical parameters, not technical restrictions and barriers. The intermittency and uncertainty of renewable energy sources could be managed, but managers in the traditional systems are not interested in new and radical technology because it might reduce their dominance over the system.

Wang and Chen (2010) examined the opportunities for and barriers to CDM in enhancing renewable energy deployment to achieve ambitious targets in China. In terms of primary energy consumption, more than 70 % of electricity is generated by coal in China, making it one of the biggest CO_2 emitters in the world. Wang and Chen concluded that there are three barriers to utilizing CDM activities in China: additional conditions (based on article 12 of the Kyoto Protocol), lower proportional certified emission reduction credit revenues, and the lack of incentives for technology transfer. Wee et al. (2012) evaluated renewable energy sources from a supply chain perspective and examined renewable energies based on supply chains, performance, barriers to RET deployment, and strategies used to overcome these barriers. They identified conversion cost, location constraints, and complex distribution networks as main barriers to developing RETs and argued that they could be removed through the participation of governments, researchers, and other stakeholders in the development of renewable energies.

Egbue and Long (2012) investigated barriers to the widespread adoption of electric vehicles through an internet-based survey and analyzed consumers' insights regarding the purchase of an EV. The results showed that consumers viewed the limitation of the battery operation of EVs as the most important concern followed by high cost and charging infrastructure. Egbue and Long also found that 25 % of respondents agreed and 43 % of respondents were neutral regarding perceptions of the sustainability of EVs relative to other vehicles. The results indicated that attitudes and knowledge regarding EVs differed according to age, gender, and education. Egbue and Long argued that although there are major technical challenges including battery, cost, and infrastructure, the most important factor, which should be taken into account by policymakers, is consumer acceptance as the key to the commercialization of EVs. The importance of social acceptance, buyer readiness, and informing customers about the benefits of new technology was emphasized in this research.

4.6 National and International Environmental Policies

Climate change is considered a global challenge, and international cooperation is required to overcome this problem. In general, environmental policy refers to the commitment of an entity at the level of an organization, government, or group of countries (regional or international) to regulations regarding environmental issues. International institutions such as the United Nations Environmental Program (UNEP), World Environment Organization (WEO), and Intergovernmental Panel on Climate Change (IPCC) for global environmental management have made several efforts in this direction. Many bilateral and regional agreements and international organizations (e.g., OECD) deal with environmental issues. Both resource-rich and resource-poor countries have tried to apply degrees of environmental protection to overcome climate change issues. The possibility of using energy transition and new technology has created challenges in terms of energy security and environmental management (Bhattacharyya 2011). A few organizations such as the International Maritime Organization (IMO), a specialized agency of the United Nations that addresses various aspects of international shipping, have the power to enforce the performance of regulations made by IMO regarding the control and prevention of marine pollution. Most policies made by organizations are performed by members based on their commitments. Because of the state-centric nature of international cooperation, states play the most important role in making international environmental regulations (Axelrod et al. 2011).

Denmark is considered a front-runner in renewable energy deployment because 33 % of its power supply is generated by renewable energy sources (20 % wind and 13 % biofuels) and it exports excess electricity generated by offshore wind turbines to neighboring countries (i.e., Norway and Sweden). On 17 June 2005, the Danish government released *Energy Strategy 2025*, which focused on initiatives for energy saving, renewable energy, climate change, energy markets, and technology. It set the long-term target of 100 % independence from fossil fuels (IEA 2012c). The Danish government applied incentive policies to achieve ambitious targets such as hydrogen-fuelled and electric cars, which will be tax free; 35 million DKK financial support for research on electric cars; and a further 25 million DKK for research on solar, wave power, and other renewable energy annually for 4 years. It released *Energy Strategy 2050* in February 2011, which outlined a number of new short- and medium-term policy incentives. Based on the IEA report, the key elements in the future Danish energy system were identified as follows: highly efficient energy consumption; electrification of heating, industry, and transport; more electricity from wind power; efficient utilization of biomass resources; utilization of biogas; PV solar modules and wave power as supplements; widespread energy-based district and individual heating; and intelligent energy system.

In Germany, renewable energy sources comprise 22 % of electricity generation. Germany has progressed during the last two decades in reducing CO_2 emissions and the energy intensity of its economy. Energy Package 2011 is constituted of the Energy Industry Act, Renewable Energies Act, Nuclear Energy Act, Energy and Climate Act, Act to Strengthen Climate-Compatible Development in Cities and Municipalities, Act on Tax Incentives for Energy-Related Modernization of Residential Buildings, and Ordinance on the Award of Public Contracts (IEA 2013). Germany has formulated its energy policy to make it a world leader in the field of energy efficiency and environmental protection. Its targets include ambitious environment protection indices as an essential part of energy policy including 40 % reduction in GHGs by 2020, 55 % by 2030, 70 % by 2040, and 80–95 % by 2050 compared to 1990 levels. In Germany, the elements of the integrated energy and climate program and GHG emission reduction target areas are as follows: reducing electricity consumption; modernizing fossil-fired power stations; promoting electricity generation from renewable energies; promoting CHP generation; modernizing building and heating systems; using renewable energies in heat production; and implementing energy-saving measures in the transport sector. The achievement of these targets would lead to a reduction of 270 Mt carbon dioxide per year by 2020. The Renewable Energy Sources Act is considered the key support instrument in generating electricity by renewable energy sources.

It has been forecast that by 2050, the Earth's population would have increased to more than 9 billion people. Without new policies and with increasing emissions caused by the combustion of fossil fuels, the global GDP will quadruple, and energy consumption is projected to grow by more than 80 %. Because no single policy is considered a solution to climate change, a mix of policy instruments is applied to reduce GHG emissions. The essential parts of this policy mix according to the OECD are the following: national climate change strategies; price-based instruments (cap and trade); carbon taxes and removing fossil fuel subsidies; command and control instruments and regulations; technology support policies including R&D; voluntary approaches; public awareness campaigns; and information tools (OECD 2012). The coverage and scope of work for each instrument vary among countries. For example, carbon pricing is the main policy in Australia, the EU, Korea, and the UK. Canada, China, France, Germany, India, Italy, Japan, Russia, and Portugal focus on policies regarding energy efficiency. Some policy tools for climate change mitigation which could be used as price-based instruments include the following: taxes on CO_2 emissions; taxes on inputs or outputs of process; removal of environmentally harmful subsidies; subsidies for emission-reducing activities; and emission trading systems.

Hirschl (2009) provided an analysis of renewable energy policies at the international level. This analysis showed that the dominant role of fossil fuels and nuclear energy supported by subsidies makes it difficult for renewable energy technologies to play an effective role in international energy and climate change policy. Hirschl argued that international organizations such as the IEA and the World Bank have provided minimum financial support to enhance renewable energy deployment and that they prefer to focus on conventional centralized energy systems. In his view, emissions trading could be developed if carbon dioxide prices stay high, but so far, this has not been sufficient. Although feed-in tariffs create better incentives for investment in renewable energy markets, they have made only small contributions to the development of renewable energy sources worldwide because of the small number of feasible CDM projects. On the other hand, policymaking within the EU seems to be influenced by political considerations.

Based on Axelrod et al. (2011), from an integrated perspective, political will and public support have played key roles in the EU's success in formulating and implementing environmental policy. While they mentioned the EU as the most advanced regional organization of states and as having a comprehensive environmental policy regime, they also noted a debate regarding whether the "EU is an intergovernmental organization dominated by the interests of individual Member States or a functional regime that represents common transactional interests and actors." In this regard, there is much evidence of conflicts among member states based on their interests. An example is the carbon dioxide standard for new cars, which created conflict between the owners of powerful automobile industries (i.e., Germany and France) and other member states.

References

Ackermann T, Andersson G, Söder L (2001) Overview of government and market driven programs for the promotion of renewable power generation. Renew Energy 22(1):197–204

Anton WRQ, Deltas G, Khanna M (2004) Incentives for environmental self-regulation and implications for environmental performance. J Environ Econ Manag 48(1):632–654

Axelrod RS, VanDeveer SD, Downie DL (2011) The global environment: institutions, law and policy. Earthscan, London

Bhattacharyya SC (2011) Energy economics: concepts, issues, markets and governance. Springer, London

Bosselmann K, Richardson BJ (1999) Environmental justice and market mechanisms: key challenges for environmental law and policy. Kluwer Law International, The Hague

Carlsson B, Jacobsson S (1997) In search of useful public policies: key lessons and issues for policy makers. Econ Sci Technol Innov 10:299–315

Costantini V, Crespi F (2008) Environmental regulation and the export dynamics of energy technologies. Ecol Econ 66(2):447–460

de Joode J, Jansen JC, van der Welle AJ, Scheepers MJJ (2009) Increasing penetration of renewable and distributed electricity generation and the need for different network regulation. Energy Policy 37(8):2907–2915. doi:10.1016/j.enpol.2009.03.014

del Río P, Bleda M (2012) Comparing the innovation effects of support schemes for renewable electricity technologies: a function of innovation approach. Energy Policy 50:272–282

Dinica V (2006) Support systems for the diffusion of renewable energy technologies—an investor perspective. Energy Policy 34(4):461–480

Dinica V (2011) Renewable electricity production costs—a framework to assist policy-makers' decisions on price support. Energy Policy 39(7):4153–4167

Dosi G (1988) Sources, procedures, and microeconomic effects of innovation. J Econ Lit 26(3):1120–1171

Egbue O, Long S (2012) Barriers to widespread adoption of electric vehicles: an analysis of consumer attitudes and perceptions. Energy Policy 48:717–729

El Fadel M, Rachid G, El-Samra R, Boutros GB, Hashisho J (2013) Knowledge management mapping and gap analysis in renewable energy: towards a sustainable framework in developing countries. Renew Sustain Energy Rev 20:576–584. doi:10.1016/j.rser.2.012.11.071

Fouquet D, Johansson TB (2008) European renewable energy policy at crossroads—focus on electricity support mechanisms. Energy Policy 36(11):4079–4092

Foxon TJ, Gross R, Chase A, Howes J, Arnall A, Anderson D (2005) UK innovation systems for new and renewable energy technologies: drivers, barriers and systems failures. Energy Policy 33(16):2123–2137. doi:10.1016/j.enpol.2004.04.011

Gunningham N, Grabosky P, Sinclair D (1998) Smart regulation: designing environmental policy. Oxford Clarendon Press, Oxford

Hirschl B (2009) International renewable energy policy—between marginalization and initial approaches. Energy Policy 37(11):4407–4416

IEA (2012a) Energy technology perspectives 2012. OECD Publishing, Paris

IEA (2012b) World energy outlook 2012. OECD Publishing, Paris

IEA (2012c) Energy policies of IEA countries: Denmark 2011. OECD Publishing, Paris

IEA (2013) Energy policies of IEA countries: Germany 2013. OECD Publishing, Paris

Jaffe AB, Newell RG, Stavins RN (2005) A tale of two market failures: technology and environmental policy. Ecol Econ 54(2):164–174

Jaffe AB, Palmer K (1997) Environmental regulation and innovation: a panel data study. Rev Econ Stat 79(4):610–619

Jaffe AB, Peterson SR, Portney PR, Stavins RN (1995) Environmental regulation and the competitiveness of US manufacturing: what does the evidence tell us? J Econ Lit 33(1):132–163

Johnson B, Gregersen B (1995) Systems of innovation and economic integration. J Ind Stud 2(2):1–18

Kemp R (1997) Environmental policy and technical change: a comparison of the technological impact of policy instruments. Edward Elgar, Cheltenham

Klein Woolthuis R, Lankhuizen M, Gilsing V (2005) A system failure framework for innovation policy design. Technovation 25(6):609–619

Lema A, Lema R (2013) Technology transfer in the clean development mechanism: insights from wind power. Glob Environ Change Hum Policy Dimens 23(1):301–313. doi:10.1016/j.gloenvcha.2012.10.010

Lewis JI, Wiser RH (2007) Fostering a renewable energy technology industry: an international comparison of wind industry policy support mechanisms. Energy Policy 35(3):1844–1857

Loock M (2012) Going beyond best technology and lowest price: on renewable energy investors' preference for service-driven business models. Energy Policy 40:21–27

Martinot E, Sawin J (2012) Renewables global status report. Renewables 2012 Global Status Report, REN21. http://www.martinot.info/REN21_GSR2012.pdf

Menanteau P, Finon D, Lamy ML (2003) Prices versus quantities: choosing policies for promoting the development of renewable energy. Energy Policy 31(8):799–812

Murphy J, Gouldson A (2000) Environmental policy and industrial innovation: integrating environment and economy through ecological modernisation. Geoforum 31(1):33–44

Nemet GF (2009) Demand-pull, technology-push, and government-led incentives for non-incremental technical change. Res Policy 38(5):700–709

Nykamp S, Andor M, Hurink JL (2012) 'Standard' incentive regulation hinders the integration of renewable energy generation. Energy Policy 47:222–237. doi:10.1016/j.enpol.2012.04.061

OECD (2011) Better policies to support eco-innovation. OECD Publishing, Paris

OECD (2012) OECD environmental outlook to 2050. OECD Publishing, Paris

Owen AD (2006) Renewable energy: externality costs as market barriers. Energy Policy 34(5):632–642. doi:10.1016/j.enpol.2005.11.017

Painuly JP (2001) Barriers to renewable energy penetration; a framework for analysis. Renew Energy 24(1):73–89

Patlitzianas KD, Doukas H, Psarras J (2006) Enhancing renewable energy in the Arab States of the Gulf: constraints & efforts. Energy Policy 34(18):3719–3726

Reddy S, Painuly JP (2004) Diffusion of renewable energy technologies – barriers and stakeholders' perspectives. Renew Energy 29(9):1431–1447

Sawin J (2004) National policy instruments: policy lessons for the advancement & diffusion of renewable energy technologies around the world. International Conference for Renewable Energies, Boon. http://www.worldfuturecouncil.org/fileadmin/user_upload/Miguel/Sawin__2004__National_policy_instruments.pdf

Smith K (1997) Economic infrastructures and innovation systems. Systems of innovation. Pinter, London

Sovacool BK (2009a) The cultural barriers to renewable energy and energy efficiency in the United States. Technol Soc 31(4):365–373

Sovacool BK (2009b) The intermittency of wind, solar, and renewable electricity generators: technical barrier or rhetorical excuse? Util Policy 17(3):288–296

Steinbach J, Ragwitz M, Bürger V, Becker L, Kranzl L, Hummel M, Müller A (2013) Analysis of harmonisation options for renewable heating support policies in the European Union. Energy Policy 59:59–70

Verbruggen A (2009) Performance evaluation of renewable energy support policies, applied on Flanders' tradable certificates system. Energy Policy 37(4):1385–1394

Wang QA, Chen Y (2010) Barriers and opportunities of using the clean development mechanism to advance renewable energy development in China. Renew Sustain Energy Rev 14(7):1989–1998. doi:10.1016/j.rser.2010.03.023

Wang Y (2006) Renewable electricity in Sweden: an analysis of policy and regulations. Energy Policy 34(10):1209–1220. doi:10.1016/j.enpol.2004.10.018

Wee HM, Yang WH, Chou CW, Padilan MV (2012) Renewable energy supply chains, performance, application barriers, and strategies for further development. Renew Sustain Energy Rev 16(8):5451–5465. doi:10.1016/j.rser.2012.06.006

Zhao ZY, Zuo JA, Fan LL, Zillante G (2011) Impacts of renewable energy regulations on the structure of power generation in China – a critical analysis. Renew Energy 36(1):24–30. doi:10.1016/j.renene.2010.05.015

Chapter 5
Financing Renewable Energy Development

5.1 Introduction

The economic view is an essential part of renewable energy deployment and its progress. Without economic advantage, renewable energy technology will not be able to compete with conventional resource technologies. On the other hand, it is difficult to establish a transparent figure for the unit cost of renewable energy compared to conventional sources. External costs such as social and environmental costs are included in conventional sources. Moreover, the subsidies paid for the consumption of fossil fuels become a barrier and make it expensive for alternative sources to compete. The aim of increasing the contribution of renewable energy in the total primary source of energy supply is of worldwide importance in mitigating the negative effects of climate change.

In reality, there is a gap between the actual share and optimal level of renewable energy consumption in the world. Furthermore, a huge amount of investment is made on conventional energy sources compared to renewable energy sources. Alternative policies for environmental protection could be applied in the form of economic incentives and nonincentive regulations. Regulatory frameworks and relative policies have been discussed already. Economic policies could give an incentive for using renewable energy or imposing taxes on emission generation or fossil fuel consumption.

Three types of support mechanisms are used widely: feed-in tariffs, tax incentives, and tradable green certificates. We do not consider direct monies paid to producers or consumers because our purpose is to apply a mechanism to encourage the creation of a renewable energy market. Direct financial transfer may lead to the enhancement of renewable energy consumption, but the main target (i.e., market creation) is not achieved. Generally, different kinds of economic instruments (e.g., capital grants, grants to infrastructure, utility procurement) are available for renewable energy technologies. However, most of these are not appropriate for electricity produced by individually distributed generators. Considering that many

© Springer Science+Business Media Singapore 2015
A. Heshmati et al., *The Development of Renewable Energy Sources and its Significance for the Environment*, DOI 10.1007/978-981-287-462-7_5

of the most promising technologies to deploy renewable energy to achieve targets for energy efficiency or carbon reduction require investment in small-scale energy production systems (such as residential buildings), these mechanisms could be used to enhance renewable energy development on the desired scale. In particular, they will be applicable to the renewable energy market, which is constituted by a large number of individual energy suppliers.

The rapid growth rate of renewable energy during recent decades has been possible through decreasing technology costs, increasing fossil fuel prices, and the continued payment of subsidies by the state. According to an IEA report in 2012, the subsidies will increase from $88 billion in 2011 to almost $240 billion in 2035 (IEA 2012). On the other hand, fossil fuel consumption subsidies were estimated at $523 billion in 2011, which is nearly six times more than the financial support allocated to renewable energy. This means that the support of conventional sources of energy overshadows the support of renewable energy sources. To highlight this and provide some ideas on reversing it, we discuss financial support mechanisms (i.e., feed-in tariffs, tax incentives, and tradable green certificates) and cross-national incentive policies for clean development mechanisms in the following sections.

5.2   Feed-in Tariff Energy Supply Policy

A feed-in tariff (FIT) is a policy used as a support mechanism to accelerate investment in renewable energy (RE) technologies. According to Couture and Gagnon (2010), "a feed-in tariff (FIT) is an energy supply policy focused on supporting the development of new renewable energy projects by offering long-term purchase agreement for the sale of RE electricity." Couture et al. (2010) pointed to three essential provisions for the success of FIT policies: guaranteed access to the grid, stable and long-term power purchase agreements, and calculation of prices on the basis of the unit costs of power generated by renewable energy sources. Technologies such as wind power are priced lower than solar PV because of the latter's higher cost. However, FIT policies could be considered as a controlling regulation because of the ability of governments to direct the market according to the rates in the contracts. An expectation of lower rates in the future could cause a rush in the market to receive the existing FIT rate. The tariffs may be used as fixed rate (higher than market price) or as a markup that is added to the current market price.

According to *Renewables Global Status Report* (GSR 2012), at least 109 countries had used some type of FIT policy by early 2012. FIT policies had been used in at least 65 countries and 27 states by 2012 (Martinot and Sawin 2012). In late 2011, Germany was successful in connecting its one millionth PV system mainly because of the low rate of FIT and the expectation that prices would continue to decrease. Based on the *GSR 2012*, this new connected PV system (around 7.5 GW) set 24.8 GW as the cumulative installed capacity, accounting for 3.1 % of Germany's power generation and almost 8 % of peak load demand. Future examples of using FIT contract rates as the stimulus for PV system installation are as follows: Italy brought 9.3 GW of PV to benefit from more advantageous rates in 2010; the UK increased its capacity to 1 GW driven by two rounds of rate reductions; France operated more than 1.6 GW by changing its FIT rates; China's market has quadrupled mostly because of the national FIT policy, increasing its cumulative installed capacity to 3.1 GW and making it the dominant player in Asia (Martinot and Sawin 2012). Figure 5.1 shows the effects of FIT policy on developing renewable energy installation in the UK. As the figure shows, PV installation has been affected widely by FIT policy.

The UK government introduced a FIT-supporting mechanism on April 1, 2010, in order to enhance small-scale renewable energy deployment and low-carbon power generation technologies. This policy covers five technologies—solar PV, wind, hydro, anaerobic digestion, and micro CHP plants. As Fig. 5.1 shows, almost 2 million kW have been installed on the basis of the FIT-supporting mechanism, and the majority of the installed capacity is from solar PV sources. Figure 5.2 shows the number of cumulative installations in the same period.

Financial support mechanisms are gaining importance in enhancing renewable energy development. Conventional sources of energy could be replaced with renewable energy technologies (RET) in order to mitigate environmental damage caused by old electricity power generation technology. Ringel (2006) examined the most common types of support systems in the EU including FITs and green certificates in order to evaluate their advantages and disadvantages in terms of effectiveness and efficiency. Based on this analysis, both mechanisms contributed

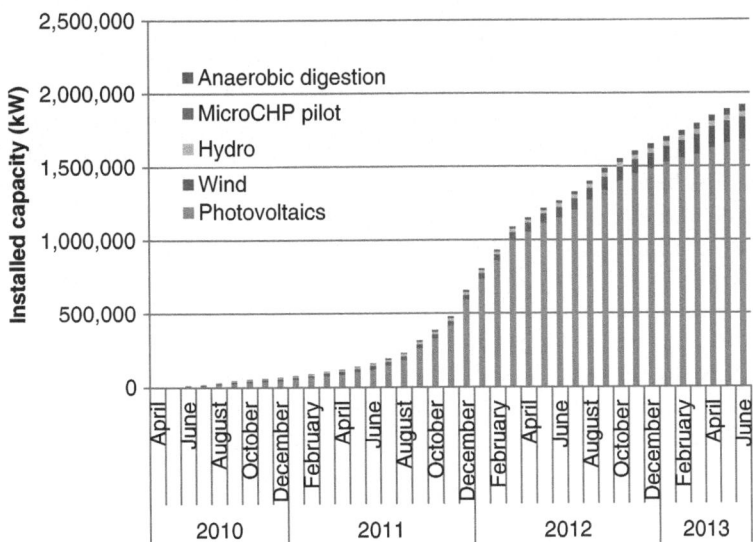

Fig. 5.1 Cumulative installed capacity of renewable energy based on FITs in the UK (Source: Department of Energy and Climate Change, UK, 2013)

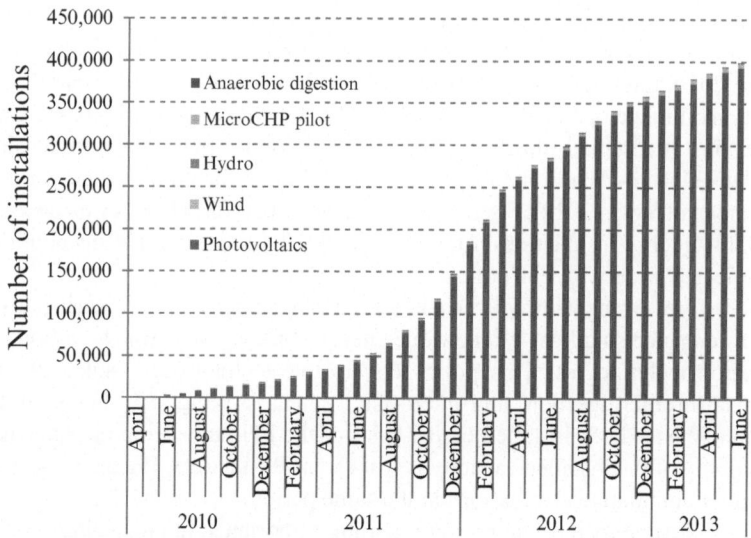

Fig. 5.2 Cumulative installed number of renewable energy based on FITs in the UK (Source: Department of Energy and Climate Change, UK, 2013)

to enhancing the share of power generation by renewable energy. However, many member states tried to shift from a feed-in system to green certificates while experiencing both systems. The results show that FITs are successful in encouraging individuals to use renewable energy sources, mainly wind energy in Denmark and Germany. However, FITs do not have enough capability to create a liberalized single electricity market. With regard to the EU market, there is an intention to promote the use of green certificates instead of FITs because of the upcoming market for emission certificates. It has been argued that the installation of a harmonized FIT system is almost impossible for reasons of feasibility and politics at the European level. It has been concluded that no single policy is capable of solving renewable energy development in a harmonized European market for electricity, so the debate regarding the application of appropriate support policies continues.

Rickerson et al. (2007) analyzed support mechanisms in Europe and evaluated their application to policies in the USA. They pointed to the debate in Europe about FITs and renewable portfolio standards (RPSs) with the aim of applying a single harmonized policy in the US market. They emphasized that US states are not under federal pressure to comply with a harmonized energy policy. Therefore, they can apply different approaches to take advantage of the renewable energy policy. Considering the experience in Europe and the emerging FIT system as a viable policy in the USA, it is expected that this policy will continue to set specific targets for renewable technologies. Some studies have held that FITs could be used for emerging technologies such as PV and RPS should be used to enhance near-market renewable energy technologies. Butler and Neuhoff (2008) compared FITs with quota and auction mechanisms to support wind energy development in the UK and Germany by surveying project developers. The result showed that "a frequent criticism of the feed-in tariff is that it does not generate sufficient competition." They concluded that deployment levels are much higher in Germany compared to the UK in terms of installed capacity, which is attributed to the limitation of procedure and cost factors in the UK.

Lesser and Su (2008) examined an economically efficient FIT structure to enhance renewable energy deployment, proposing a two-part FIT system that used both capacity- and market-based payments to achieve targets. Based on the finding of this research, the following actions should be performed in a two-part FIT: define technologies that are eligible to receive subsidies, set a desired capacity goal for each technology, establish a contract period for each FIT, and set a payment period for the winning auction prices. They argued that the proposed two-part FIT system could be able to eliminate cost errors caused by difficulties faced from using the system based on the Public Utility Regulatory Policies Act of 1978 (PURPA). This Act required utilities to buy electricity from independent renewable energy and cogeneration plants and proved to be a controversial policy. Under PURPA, the purchase price was based on avoided cost and was left to individual states (Rickerson and Grace 2007). This Act caused some difficulties because it imposed an expensive burden on the utility rate paid by consumers. Lesser and Su (2008) indicated that a two-part FIT policy would create competition and lead to long-term financial stability, thus avoiding overestimates or underestimates in payments to RET developers.

Solano-Peralta et al. (2009) proposed a FIT scheme to take advantage of renewable energy using conventional sources as a hybrid system in isolated areas. They introduced a model that integrated a photovoltaic (PV) generator into a diesel system for off-grid regions in Ecuador. They compared the results with the existing diesel system. They proposed that the Renewable Energy Premium Tariff (RPT) plan is an alternative mechanism to deploy renewable energy in off-grid regions. The results showed that the PV–diesel hybrid system is more cost effective than the stand-alone diesel system. They found that if the optimistic lower value in terms of life-cycle cost is considered for PV–diesel hybrid generators, the RPT value for a model with a neutral net present value (NPV = 0) is estimated at 0.59 and 0.57 $/kWh for 62 and 79 % PV fraction scenarios compared to the FIT values of 0.52 $/kWh paid for PVs installed in the mainland of Ecuador. This scheme could be applied to other renewable energy technologies. Hence, it could assist in the introduction of RETs as a sustainable energy option for residents located in isolated off-grid areas.

Couture and Gagnon (2010) analyzed a variety of feed-in tariff payment models for power generated by renewable energy sources in order to determine the advantages and disadvantages of different FIT models based on dependency on the market price of electricity. They argued that the overall rate of renewable energy enhancement is influenced by the FIT payment structure and its impact on investor risk. They found that market-independent policies are stronger and more cost effective than market-dependent options because of their greater investment security and lower cost of renewable energy deployment. This condition could lead to lower risk and enable investors to have a predictable cash flow. It is also attractive to small investors and community-based projects because of purchase obligations and lower barriers to entry. Therefore, the advantages associated with market-independent policies are that they are able to increase public support, which may lead to greater participation of individuals and higher levels of renewable energy penetration into the electricity market.

Rigter and Vidican (2010) developed an equation for the cost of PV to calculate the optimal FIT that must be applied in China in order to achieve its ambitious target to expand the share of electricity generated by solar power. Their calculation for a program starting in 2010 for 25 years with an assumption of applying 5 % for discount rate and 2 % for increasing the rate of electricity per year estimated that the government is charged almost 3.13$ for the installation of 1 watt of small-scale solar energy. The figure will change to 2.45$ if the program is postponed to the end of 2011. This calculation shows why the Chinese government is hesitant about the starting date of the program. Rigter and Vidican argued that the FIT policy is able to compensate the negative effects of low electricity prices on the feasibility of solar PV. Moreover, these costs could decrease rapidly over time because of innovation and technology development.

Wand and Leuthold (2011) employed a partial equilibrium approach to analyze the potential effects of a FIT policy in Germany for small rooftop solar PV systems installed between 2009 and 2030. They applied a dynamic optimization model that considered learning-by-doing, technology diffusion, and yield-dependent demand in order to maximize social welfare. Wand and Leuthold calculated a wide range of effects on social welfare including the net social cost of 2.014 billion Euros under the Business as Usual (BAU) scenario and net benefits of 7.586 billion Euros from the benefits of solar PV deployment. They argued that the BAU scenario underestimated actual prices, while the positive scenario reflected FIT policies established recently in Germany for the residential PV market. We should also take into account that the negative welfare effect is viewed differently by policymakers in Germany because of its dependency on fossil fuels and concerns for a secure energy supply in the future. Krajacic et al. (2011) examined the FIT system used to implement energy storage technologies to overcome the intermittency of renewable energy sources and technical capability of power networks. They analyzed FIT applications for pumped hydro storage (PHS), hybrid wind–hydro pumped storage (WHPS), a hydrogen storage system (HSS), PV and batteries, and desalination systems. They also evaluated the FIT mechanism to apply these technologies in some islands and outer areas. After the successful application in these areas, they argued that energy storage tariffs could be applied in main power systems to enhance energy storage usage and to optimize existing utilities in the market.

In a recent study, Schallenberg-Rodriguez and Haas (2012) evaluated two alternative options of FIT—fixed and premium—and examined the evolution of the implementation of both mechanisms in Spain. These mechanisms, which are employed in Spain, have led to successful renewable energy deployment for power generation. The results showed that the direct burden of customers generated by a fixed FIT policy is smaller than that imposed by the premium and it provided higher security for investors compared to the premium option. On the other hand, the premium mechanism is market oriented and provides a slightly higher income, which could be more attractive for investors. Schallenberg and Haas stated that the premium mechanism might lead to overcompensation when prices increase quickly but this problem could be avoided by setting a cap value. The findings indicated that a fixed FIT policy should be used for technologies that have not achieved high market penetration and in cases that require building a market environment.

In Portugal, FITs have been the main policy used to enhance power generated by renewable energy sources. Proenca and St Aubyn (2013) simulated a model to evaluate economic and environmental impacts of a FIT policy in Portugal. They employed a hybrid top–down/bottom–up general equilibrium model to analyze the interaction between energy, economics, and environmental issues in relation to energy policies. The results showed that the FIT policy provided an effective and cost-efficient instrument to promote renewable energy sources for electricity generation. In addition, economic adjustment cost was low, and the deployment of renewable energy led to significant reductions in CO_2 emissions. Jenner et al. (2013) estimated a panel data model over the years 1992–2008 to analyze the effectiveness of FIT policies for developing solar PVs and onshore wind turbines in EU-26. They applied a fixed-effect model at the country level and introduced a measure of the return on investment (ROI) to evaluate the strength of the policy. The results showed that solar PV development in Europe has been driven by FIT policies through their impact on ROI for investors. The results also implied that the combination of FIT policy, electricity price, and unit cost served as a more accurate determinant for power generation by renewable energy sources than a stand-alone policy measure did.

A summary of empirical research conducted on feed-in tariff policy along with their key findings are provided in Table 5.1. The subjects include mainly design of FIT, comparing FIT with other alternative policies, as well as its effects on promoting renewable energy. The results point to specific effects of FIT policy considering its diverse technology innovation, environmental, economic, and social impacts.

5.3 Tax Incentive Policies

Tax exemption is used as a fiscal incentive measure to enhance renewable energy deployment in many countries. Tax credits could be applied for the investment, production, or consumption segments of electricity generated by renewable energy sources. Policies aimed at encouraging renewable energy consumption could apply tax credits on the purchase and installation of renewable equipment to facilitate the penetration of renewable energy deployment into the market. For example, a draft bill was introduced in Poland for a renewable heat obligation in private and public buildings and to provide a tax credit for private customers for solar thermal energy

Table 5.1 Some empirical studies regarding feed-in-tariff policies

Author	Subject	Result
Ringel (2006)	Fostering use of renewable energies in EU, comparing FIT and green certificate	There is an intention to promote green certificate instead of FITs due to the upcoming market for emission certificates. Using a harmonized FIT is almost impossible in Europe.
Rickerson et al. (2007)	Using FIT to meet US renewable targets	The US can apply different approaches to take advantage of the renewable energy policy. FIT could be used for emerging technologies and RPS should be used to enhance near-market energy technologies.
Butler and Neuhoff (2008)	Comparison of FIT, quota, and auction to support wind power	They noted a very low level of competition at the operational stage for all three funding schemes. Deployment levels are much higher in Germany compared to the UK.
Lesser and Su (2008)	Design of an economically FIT structure	They proposed a two-part FIT system constituted of both capacity and market-based payment to achieve targets.
Solano-Peralta et al. (2009)	Custom-made support scheme for hybrid systems in isolated regions	They proposed Renewable Energy Premium Tariff (RPT) plan for off-grid regions. RPT value for a model with NPV = 0 is estimated at 0.59 and 0.57 $/kWh for 62 and 79 % PV fraction scenarios, compared to 0.52 $/kWh FIT values paid for PV installed in the mainland of Ecuador.
Couture and Gagnon (2010)	Analysis of FIT remuneration models	Market-independent policies provide a stronger and cost-effective policy compared to market-dependent options due to greater investment security and lower cost for renewable energy deployment.
Rigter and Vidican (2010)	Cost and optimal FIT for small-scale PV in China	FIT policy is able to compensate negative impacts of low electricity prices on feasibility of solar PV, and these costs could be rapidly decreased over the time as a result of innovation and technology development.

Wand and Leuthold (2011)	FIT for PV in Germany	They found a wide range of impacts on social welfare, from net social cost of 2.014 billion Euros under a Business As Usual (BAU) scenario to net benefits of 7.586 billion Euros under a positive perspective of solar PV deployment.
Krajacic et al. (2011)	FIT for promotion of energy storage technologies	Energy storage tariffs could be applied in main power systems to enhance energy storage usage and to optimize existing utilities in the market.
Schallenberg-Rodriguez and Haas (2012)	Fixed FIT versus premium in Spain	Fixed FIT policy should be used for those technologies that have not achieved a high market penetration and where it is required to build a market environment.
Proenca and St Aubyn (2013)	Effects of FIT to promote renewable energy in Portugal	FIT policies provided an effective and cost-efficient instrument to promote renewable energy sources for electricity generation. Deployment of renewable energy led to significant CO_2 emissions reduction.
Jenner et al. (2013)	Assessing the strength and effectiveness of FIT in the EU	Combination of policy with electricity price and unit cost are more important to serve as a determinant for power generation by renewable energy sources than a stand-alone policy measure.

(Martinot and Sawin 2012). Tax policy is also used as a useful instrument to reduce fossil fuel consumption. A carbon tax imposed by a government imposes a higher cost burden for burning fossil fuels and increases investment in renewable energy sources. Demand for energy could be influenced by carbon tax through the relative cost of using different fuels. Following the Fukushima Daiichi accident, Japan set a target to triple power generation from renewable energy by 2030 over that generated in 2010. In this regard, Japan released a new FIT system to support renewable energy and employ other subsidy mechanisms such as tax credits, investment grants, and loans (IEA 2012).

Using tax credits as a supportive policy in some countries such as Germany could be applied in other countries, especially because the lack of competitiveness in conventional sources of energy is a typical barrier for most countries (Gutermuth 1998). Exemption of fuels produced by renewable energy sources (RES) such as biomasses from other taxes was applied by Germany, which made it competitive with the highly taxed conventional diesel fuel. Consequently, sales rose from 40,000 t in 1996 to 100,000 t in 1997. A similar policy was applied to electric vehicles exempting EVs from motor vehicle tax for 5 years. The main reason that this instrument is attractive is that it makes cash available. Therefore, it could be an important financial incentive for private investors and an opportunity to make small investments because it directly increases investor liquidity. Many economists believe that employing a carbon dioxide emission tax or emission trading mechanism is considered a good policy to mitigate emissions at the lowest cost (Kalkuhl et al. 2013).

Kahn and Goldman (1987) analyzed the sensitivity of renewable energy and the cogeneration project's internal rate of return (IRR) based on tax changes created by the US federal tax code according to PURPA. The electricity market was stimulated by tax incentives under PURPA. Kahn and Goldman investigated six different technologies to evaluate the effects of tax reform on the financial viability of projects. Based on their calculations, capital-intensive projects such as wind turbine, small hydro, geothermal, and wood-fired electricity were not financially viable because of the expiration of energy tax credits (ETCs). The results showed that gas-fired cogeneration technology is the most beneficial under the tax reform. Walsh (1989) applied a two-period utility maximization model of household behavior to investigate the relationship between federal and state tax credits and energy saving improvement in the USA. The federal government reduced income tax liability by 15 % of expenditure for conservation facilities from 1977 to 1986. The results indicated that tax credits are not an effective solution to subsidizing energy conservation activities. This could be because of the small discount rate applied by the tax reform, the inconvenience of claiming the credit, or lack of knowledge about the effects of price reduction.

Alfsen et al. (1995) simulated a model to assess the possible effects of carbon tax in Western Europe and the partial deregulation of the power generation industry on CO_2, SO_2, and NO_x emissions. They assumed two regimes, plan-efficient and cost-efficient, in order to examine the effectiveness of carbon tax. They found that emissions were reduced by European Community (EC) tax by 6–10 % under both regimes compared to a scenario without the tax. The regime change from plan-efficient to cost-efficient caused a reduction by 3 % in CO_2 and NO_x emissions, while SO_2 was reduced by 13 %. The results showed that an economic instrument for CO_2 emission reduction is able to influence the emissions of SO_2 and NO_x. Hassett and Metcalf (1995) estimated a panel data model of individual tax returns and variations in state tax policy in order to evaluate the impact of tax policies on investment in residential energy conservation in the USA. They concluded that a point change of 10 % in the rate of tax incentives for energy investment could increase the probability of investment by 24 %. Kahn (1996) examined the impact of tax credits set by the US government based on the Energy Policy Act (EPAct) of 1992 to enhance renewable energy technologies. The results showed that production tax credit was ineffective in providing incentives to invest in wind turbine power plants because it raised the financing cost. Therefore, the tax credit might not be able to reduce the unit cost of electricity generated by wind turbines compared to the nonincentive plan. Consequently, the effect of the production tax credit could be minimal.

Brännlund and Nordström (2004) formulated an econometric model of non-durable consumer demand in Sweden to analyze consumer response and welfare effects of changing energy or environmental policy. The results showed that households living in populated areas were more influenced by CO_2 tax, indicating that carbon tax has a regional distribution effect. In terms of environmental effects, petrol demand decreased by almost 11 %, affecting CO_2 emissions. The findings showed that the distribution of tax burden and welfare loss was uneven because the tax revenue was not returned at an even rate. Barradale (2010) examined the impact of repeated expiration and a short-term renewal federal production tax credit (PTC) on wind power investment. The federal tax revision created an on–off pattern, which led to a boom–bust cycle in wind power plant investment in the USA leading to a severe downtrend for investment because of the high cost of ramp-up and ramp-down. The results implied that wind power is not feasible in the absence of PTC. However, Barradale found that it was formed because of the negotiation dynamics in the power purchase agreement in the face of PTC uncertainty. Therefore, an incentive instrument applied for enhancing renewable energy production may be changed to a disincentive form if it causes uncertainty.

Galinato and Yoder (2010) introduced an integrated tax subsidy policy for CO_2 emission reduction, which is considered a compromise between the standard Pigovian tax and a traditional indirect subsidy. They argued that environmental taxes on energy are not popular politically because the taxes imposed on fossil fuels could lead to higher energy prices. Moreover, subsidies paid for renewable energy fuels are funded mainly by labor or income taxes. Galinato and Yoder suggested that revenues from carbon taxes could be used to fund subsidies for the production of low-emitter fuels. Therefore, the carbon tax and subsidy mechanism are revenue neutral within the energy industry. Levin et al. (2011) developed an energy optimization model and applied it to investigate the effects of the Renewable Energy Standard (RES) and carbon tax in the state of Georgia in the USA. The results showed that power generated from biomass cogeneration at coal plants could be considered a low-cost option but the potential was limited. The findings indicated that constant carbon tax till 2030 would lead to the replacement of coal with natural gas instead of generating electricity from renewable energy sources. Based on the calculation, low-cost biomass had the lowest levelized costs of electricity in all scenarios. The importance of a properly designed policy in order to achieve targets was emphasized in this research.

Pablo-Romero et al. (2013) investigated a variety of instruments such as tax incentives applied in Spain to enhance solar thermal energy deployment. They argued that tax incentives have not had a sufficient impact on solar thermal energy deployment because of regulatory changes at the national level, which complicated the system. It has also been stated that tax incentives would be more effective if they were associated with a proper financing mechanism. In this regard, Lehmann (2013) employed an analytic partial equilibrium model of the electricity sector to show that an optimal policy to mitigate climate change could be designed efficiently by applying an emission tax along with a subsidy for renewable energy sources. Lehmann analyzed the performance of revenue-neutral fixed tariff and relevant differences compared to a government-funded premium tariff. He argued that a premium tariff policy may be better than a noncontinuous adoption but some important parameters such as increasing investment uncertainty and transaction costs, concerns associated with funding tariffs from public budgets, and breaking EU completion law (because they are classified as national subsidies) should be taken into account. It has been indicated that an optimal tax rate should be less than the Pigovian level differentiated across fossil fuels and modified continuously based on technological change. This paper implies that in the presence of learning-by-doing for renewable energy technologies, a combination of emission tax and a subsidy that is modified continuously could be considered an optimal strategy.

A summary of selected empirical studies on effects of tax incentive polices and their key findings is reported in Table 5.2. The studies in general focus on the impact of various energy taxes on energy consumption and its efficient use. The results show evidence of various tax reforms on energy production, energy use efficiency, and emission reductions.

5.4   Renewable Portfolio Standard Policy

As previously mentioned, governments use a variety of policies to promote renewable energy. The renewable portfolio standard (RPS) is one of the most common policies used with FITs. In contrast to the FIT policy, which is price based (fixed-price and premium-price), the RPS policy is quantity based. This instrument requires companies to increase the amount of power generated by renewable energy sources. The RPS mechanism obligates utility companies to generate a specified share of their electricity by renewable energy. By doing so, they receive tradable certificates for every unit (e.g., 1 MWh) of power generated—these are called tradable green certificates (TGCs). Unlike the FIT mechanism in which the

Table 5.2 Some empirical studies regarding tax incentive policies

Author	Subject	Result
Kahn and Goldman (1987)	Impact of tax reform on renewable and co-generation projects	Capital intensive projects such as wind turbine, small hydro, geothermal and wood-fired electricity were not financially viable with the expiration of energy tax credits. The avoided cost price is important to develop projects.
Walsh (1989)	Energy tax credit and housing improvement	Tax credits are not considered as an effective solution to subsidize energy conservation activities. It could be due to small discount rate, uncomfortable procedure to claim the credit or lack of knowledge about price impact.
Alfsen et al. (1995)	Impacts of EC carbon tax and deregulating power supply on emissions	Emissions are reduced by European Community (EC) tax by 6–10 % under both regimes (plan- and cost-efficient) compared to the scenario without tax.
Hassett and Metcalf (1995)	Energy tax credits and residential conservation	They found that a 10 percentage point change in the rate of tax incentive for energy investment could increase the probability of investment by 24 %.
Kahn (1996)	Production tax credit for wind turbine power plants	Production tax credit is considered as an ineffective incentive for wind turbine power plants because it raises the financing cost.
Brännlund and Nordström (2004)	Carbon tax simulation using a household model	Households living in populated areas are influenced more by CO_2 tax. Petrol demand will decrease by almost 11 %, which effects CO_2 emissions.
Barradale (2010)	Wind power and the production tax credit	Wind power is not feasible in the lack of production tax credit. An incentive instrument applied for enhancing renewable energy production may be changed to a disincentive form if it comes with uncertainty.
Galinato and Yoder (2010)	An integrated tax subsidy policy for CO_2 emissions reduction	Revenues made by carbon taxes could be used to fund subsidies for low-emitter fuels. Therefore, carbon tax and subsidy mechanisms are revenue neutral within the energy industry.
Levin et al. (2011)	The role of renewable electricity credits and carbon taxes	Low-cost biomass has the lowest levelized costs of electricity in all scenarios. The importance of properly designing policy in order to achieve targets has been emphasized.
Pablo-Romero et al. (2013)	Incentives to promote solar thermal energy in Spain	Tax incentives have not had a sufficient impact on solar thermal energy deployment due to regulatory changes at the national level, which caused the system to be confused.
Lehmann (2013)	Supplementing emission tax with an FIT for renewable electricity	Optimal tax rate should be less than the Pigouvian level, differentiated across the fossil fuels, and modified continuously based on technological change.

government guarantees the purchase of generated electricity, the RPS mechanism relies on the private market for its implementation. Therefore, there is more price competition across different types of renewable energy technologies. By early 2010, RPS was in place in 56 states, provinces, and countries including more than half of the US states (Martinot and Sawin 2012). RPS policy is usually associated with the certificate trading mechanism. In the USA, a credit multiplier is applied to promote specific types of renewable energy technology. For instance, a wind multiplier means that 1 MWh of power generated by wind technology could equal three certificates for the generator (M. J. Beck Consulting 2009). Therefore, governments could use the credit multiplier as an instrument to direct revenue, investment, and job creation to a particular type of renewable energy technology.

The EU has experience with both FIT and RPS mechanisms. However, the former has led to the rapid expansion of the capacity of renewable energy and thus has been employed more than the latter. Experience with FIT is limited in the USA, which has focused more on RPS (Rickerson and Grace 2007). Although RPS policies have diffused rapidly across the USA, FIT policy is gaining in attractiveness to policymakers because of its success in the EU, particularly Germany. As of June 2010, the RPS mechanism had been applied by 29 states in the USA. Another seven states had established nonbinding renewable energy goals. Almost 65 % of the total wind capacity additions from 2001 to 2007 in the USA were motivated by state RPS policies (Wiser 2008). Some states have experienced rapid renewable energy expansion by these policies, and Texas achieved its 2015 RPS target of 5 GW installed capacity 6 years earlier than scheduled (Edenhofer et al. 2011). According to *GSR 2012*, by 2010 quotas or RPS were used in 69 states, provinces, and countries. Two additional countries applied this policy, making a total of 71 countries in 2011 compared to 92 indicated by FIT policies (Martinot and Sawin 2012). In South Korea, a FIT policy that was operational through 2011 was changed to an RPS policy in 2012. Considering that RPS policies are usually associated with trading certificates, in 2011 India launched a new Renewable Energy Certificate (REC) mechanism linked to the current quota system.

Espey (2001) discussed a main support mechanism that was introduced to promote renewable energy deployment and examined the possible effects of applying RPS based on theoretical concepts and practical evidences. Espey argued that the RPS could not be considered a stand-alone solution that would enhance renewable energy sources. However, the advantages of the RPS make it a good starting point in the transition to an international trading system (as proposed in the Kyoto Protocol) if a well-designed mechanism is employed. Berry and Jaccard (2001) analyzed the implementation of RPS in three European countries (the Netherlands, Denmark, and Italy), nine US states, and Australia. They concluded that the RPS is generally applied to generators instead of end users. Moreover, the RPS system is managed by the government in Europe but is administered by a delegation of government and independent utility regulators in the USA and Australia. Lauber (2004) compared FIT and RPS mechanisms as two options for a harmonized community framework. Lauber argued that these systems could not be measured by a common standard because they have different purposes. The FIT mechanism is appropriate to support renewable technology development and equipment industry, whereas the RPS system is more suitable in the phase of near-market competitiveness than in the early stage of technology development.

Wiser et al. (2005) examined experiences with RPS design, application, and effects in 13 US states. The results showed that the RPS performed successfully in Texas, Iowa, and Minnesota. However, it was not effective because the policies were poorly designed. Some critical design pitfalls were experienced by states such as the following: "narrow applicability, poorly balanced supply-demand conditions, insufficient duration and stability of targets, insufficient enforcement, and poorly defined or non-existent contracting standards and cost recovery mechanisms for regulated utilities and providers of last resort."

Nishio and Asano (2006) applied a quantitative analysis to evaluate the supply amount and marginal prices of electricity generated by renewable energy sources under the RPS mechanism in Japan. The results showed that the majority of power supplied under RPS was generated by wind and biomass energy. In addition, the purchase of certificates from generators in other regions and trading among retailers enabled the RPS system to be implemented effectively. Based on the analysis of the dynamic supply curve of certificates, the marginal price would increase according to the amount of electricity supplied. Kydes (2007) analyzed the effects of applying a federal 20 % nonhydropower RPS on US energy markets by 2020. The calculation showed that this policy would be effective in enhancing renewable energy technologies and reducing emissions of NO_x by 6 %, mercury by 4 %, and CO_2 by 16.5 %. It was estimated that the total electricity cost for customers would increase by almost 3 %, making a significant cost increase of 35–60 billion dollars for the power generation industry by 2020.

Carley (2009) examined the causal effects of RPS policies in states in the USA in terms of the percentage of renewable energy (RE) deployment. They applied a standard fixed model to evaluate the effectiveness of state energy programs from 1998 to 2006. The results showed significant potential for RPS policies, and an increase in the number of states that implemented this policy could lead to an increase in RE deployment. The results indicated that the percentages of RE generation were lower in deregulated states than in regulated states, implying that competitive markets stimulate RE investment. Yin and Powers (2010) applied a panel data model to determine the impact of state-level RPS policies in the USA and introduced an index to measure RPS stringency. Their findings showed that RPS policies significantly affected in-state renewable energy development. The results also showed that allowing out-of-state certificate trading had a negative impact on the effectiveness of RPS deployment. Moon et al. (2011) performed an economic analysis of biomass power generated by two technologies under an RPS scheme in order to compare the impact of biomass on combined heat and power (CHP) system capacity. The Korean government introduced a FIT program in 2006 but changed it to RPS because it failed to achieve targets in recent years. The results showed that considering current infrastructure and technological levels, biomass gasification in CHP ranging from 0.5 to 5 MW_e could be considered a good starting point for initiating RPS mechanism.

Buckman (2011) investigated the effectiveness of banding and carve-outs as two modifications of the RPS mechanism used to support high-cost types of renewable energy technologies. The UK and Italy used these modifications as particular examples of analysis banding. The USA is selected for carve-out devices. The results showed that both methods have strengths and weaknesses and could enable markets to enhance renewable energy deployment. They concluded that banding might be better than carve-outs to support high-cost renewable energy technologies. Dong (2012) examined the effectiveness of FIT and RPS to enhance wind power generation by applying a panel data analysis of 53 countries for 5 years from 2005 to 2009. The results showed that FIT policies increased power capacity by 1,800 MW more than RPS mechanisms did. This capacity would increase to 2,000 MW when the timing of policy was taken into account because the FIT system started earlier than the RPS mechanism. There was no significant difference between FIT and RPS in terms of annual capacity. Fagiani et al. (2013) simulated a model for the period 2012–2050 in order to analyze the effects of investors' risk aversion driven by profit maximization on FIT and the certificate market system. The results showed that although FIT policy could achieve better economic efficiency than RPS could, it strongly depended on regulators' decisions. In contrast, RPSs showed better performance compared to FITs in terms of cost efficiency when the degree of risk aversion was moderate.

The comparison of FIT and RPS policies showed that the former was good when a good policy to develop renewable energy sources with a low level of risk for investors is required. However, RPS is appropriate when a market view policy is supposed to be applied by the government. Europe intends to organize a single harmonized FIT system, though it is impossible because of different policies across countries in the EU. The RPS system has not been developed in Europe because most European countries use the FIT system. Hence, it seems that FIT polices are suitable to encourage developing renewable energy sources. However, RPS systems should be applied after the implementation of renewable energy sources at a certain level.

A summary of selected empirical studies investigating the effects of renewable portfolio standard policies together with their key findings is reported in Table 5.3. The key subjects of empirical studies involve design and implementation of RPS for trade, RPS effectiveness, comparisons of RPS and FIT policies, and the impact of RPS on renewable energy generation. The findings suggest that RPS is not a stand-alone solution but enhances efficiency. A number of pitfalls are identified and remedies suggested.

5.5   Cross-Country Public Incentive Policies

Cross-country incentive policies could be studied from different points of view including the transfer of climate change mitigation technologies, emission-trading schemes, and the clean development mechanism (CDM) derived from the Kyoto

Table 5.3 Some empirical results regarding renewable portfolio standard policies

Author	Subject	Result
Espey (2001)	RPS for trade with electricity from renewable energy sources	It cannot be considered as a stand-alone solution for enhancing renewable energy sources, but it is a good starting point for a transition to international trading systems.
Berry and Jaccard (2001)	Design consideration and implementation of RPS	The RPS system is managed by the government in Europe, but administered by a delegation of government and independent utility regulators in the US and Australia. It is applied alongside other support mechanisms.
Lauber (2004)	FIT and RPS options for a harmonized community framework	FIT mechanism is an appropriate policy to support renewable technology development and equipment industry, while the RPS system is more suitable in the phase of near-market competitiveness.
Wiser et al. (2005)	Evaluating experience with renewable RPS in US	Some critical design pitfalls are as follows: narrow applicability, poorly balanced supply–demand conditions, insufficient duration and stability of targets, insufficient enforcement, and poorly defined contracting standards and cost recovery mechanisms.
Nishio and Asano (2006)	Supply amount and marginal price of renewable electricity under the RPS in Japan	The majority of power supplied under RPS is generated by wind and biomass power. Purchase of certificates from generators in other regions and trading among retailers enable the RPS system to be implemented more effectively.
Kydes (2007)	Impacts of renewable RPS on the US energy markets	It will be effective to enhance renewable energy technologies alongside an emission reduction of NO_x by 6 %, mercury by 4 %, and CO_2 by 16.5 %.
Carley (2009)	State renewable energy electricity policy in the US	There is a significant potential for RPS policies and an increase in the percentage of states implementing this policy could lead to an increase in RE deployment. Percentages of RE generation are lower in deregulated states than regulated states.
Yin and Powers (2010)	State RPS promotes in-state renewable generation	RPS policies have affected in-state renewable energy development significantly. Also, allowing out-of-state certificate trading has a negative impact on RPS deployment effectiveness.

(continued)

Table 5.3 (continued)

Author	Subject	Result
Moon et al. (2011)	Economic analysis of biomass power with RPS in Korea	Considering current infrastructure and technological levels, biomass gasification CHP ranging from 0.5 to 5 MW$_e$ could be considered as a good starting point to initiate the RPS mechanism.
Buckman (2011)	The effectiveness of RPS banding and carve-outs in supporting renewable	Both methods have different strengths and weaknesses, but they could enable markets to enhance renewable energy deployment. Banding may be better than carve-outs to support high-cost renewable energy technologies.
Dong (2012)	FIT vs. RPS: An empirical test of the relative effectiveness of RPS and FIT	The FIT policy increased power capacity by 1,800 MW than the RPS mechanism. This capacity will be increased to 2,000 MW when the timing of policy is taking into account due to the fact that the FIT system was started earlier than the RPS mechanism.
Fagiani et al. (2013)	Cost-efficiency and effectiveness of renewable energy support schemes	FIT could achieve economic efficiency better than RPS, but it depends on regulators' decisions. RPS has a better performance compared to FIT in terms of cost-efficiency when the degree of risk aversion is moderate.

Protocol. Given the considerable scale of effort required to reduce GHGs, it is almost impossible for all countries to produce environmentally friendly technologies by themselves. The OECD and the World Bank have shown that climate mitigation technology trading could be affected by nontariff measures, which would also help them to promote clean technologies (Tébar Less and McMillan 2005). Trade barriers are not the only barriers to commodity movement. Total technology is constituted by the knowledge, skills, and services associated with installation and operation. Steenblik and Kim (2009) investigated the effects of tariff and nontariff barriers on trading a selection of carbon change mitigation technologies (CCMTs) identified by the Intergovernmental Panel on Climate Change (IPCC) and IEA. These included combined heat and power, district heating and cooling, solar heating and cooling, and energy efficiency motor systems. They concluded that it is necessary for tech- nology importers to review their policies in order to facilitate the diffusion of CCMT technologies made and developed in OECD countries. Adequate environmental regulations, removal or reduction of trade barriers, adequate intellectual property rights regimes, and appropriate financing mechanisms were considered incentives to transfer renewable energy technology across countries (Tébar Less and McMillan 2005).

The Kyoto Protocol introduced three mechanisms to mitigate GHG emissions: emission trading, CDM, and joint implementation (JI). Emission trading is based on an allowance transition that enables a country (listed in Annex B) to trade emission permits. In contrast, CDM and JI are classified as project based. The European Union Emission Trading Scheme (EU ETS) is an essential part of EU climate change policy. The allowance mechanism in ETS is based on three methods: grandfathering, benchmarking, and auctioning (Koh 2010). In this system, which was applied from 2005 to 2012, nearly all permits were grandfathered (Bernard and Vielle 2009). Based on the mechanism applied to mitigate emissions, incentives could be different. In the case of emission trading, the renewable energy certificate mechanism derived from RPS has gained importance. As previously, the EU aims to harmonize support mechanisms in order to facilitate market creation for trading certificates across member nations. A CDM or JI is similar to an investment project because it is able to earn both financial returns and carbon credits. Carbon credits have monetary value, and net financial returns will be affected by this value.

Carbon credit transactions enable host countries to receive significant amounts of foreign investment. CDM is specifically designed as a mechanism to channel foreign investment into non-Annex I countries (Koh 2010) Therefore, the incentives for using CDM facilities across countries are similar to incentives to attract foreign direct investment (FDI). For example, it is important to consider any regulation that may limit the inflow of FDI such as restrictions on profit repatriation of investors. Some countries may offer investment incentives and tax concessions to promote CDM projects. Krey (2004) examined 15 unilateral potential CDM projects in India and found that average transaction costs were estimated at 0.06–0.47 $/tCO$_2$ equivalent, which corresponds to approximately 76–88 % of the total transaction costs of the projects. Chaurey and Kandpal (2009) analyzed the carbon abatement potential of solar home systems (SHSs) in India and estimated that a bundled project of 20,000 SHSs could return almost Rs 1.9 million annually at 10 $/tCO$_2$ after expenses of 20 % for preimplementation transaction costs. Of course, the balance between demand and supply should be taken into account. The total demand in the Kyoto Protocol from 2008 to 2012 was estimated at 1,222 MtCO$_2$, whereas the supply was over 3,000 MtCO$_2$. Additionally, the Green Investment Scheme developed by Russia and East European countries would bring a return of more than 1,800 MtCO$_2$ (Bhattacharyya 2011). On the other hand, these numbers demonstrate the potential capacity of CO$_2$ emission reduction in several countries. By other means, there would be a large capacity available to reduce emissions if a well-designed support mechanism were in place.

References

Alfsen KH, Birkelund H, Aaserud M (1995) Impacts of an EC carbon/energy tax and deregulating thermal power supply on CO2, SO2 and NOx emissions. Environ Resour Econ 5(2):165–189

Barradale MJ (2010) Impact of public policy uncertainty on renewable energy investment: wind power and the production tax credit. Energy Policy 38(12):7698–7709. doi:10.1016/j.enpol.2010.08.021

Bernard A, Vielle M (2009) Assessment of European Union transition scenarios with a special focus on the issue of carbon leakage. Energy Economics 31:S274–S284

Berry T, Jaccard M (2001) The renewable portfolio standard: design considerations and an implementation survey. Energy Policy 29(4):263–277. doi:10.1016/s0301-4215(00)00126-9

Bhattacharyya SC (2011) Energy economics: concepts, issues, markets and governance. Springer, London

Brännlund R, Nordström J (2004) Carbon tax simulations using a household demand model. Eur Econ Rev 48(1):211–233

Buckman G (2011) The effectiveness of Renewable Portfolio Standard banding and carve-outs in supporting high-cost types of renewable electricity. Energy Policy 39(7):4105–4114. doi:10.1016/j.enpol.2011.03.075

Butler L, Neuhoff K (2008) Comparison of feed-in tariff, quota and auction mechanisms to support wind power development. Renew Energy 33(8):1854–1867. doi:10.1016/j.renene.2007.10.008

Carley S (2009) State renewable energy electricity policies: an empirical evaluation of effectiveness. Energy Policy 37(8):3071–3081

Chaurey A, Kandpal T (2009) Carbon abatement potential of solar home systems in India and their cost reduction due to carbon finance. Energy Policy 37(1):115–125

Couture T, Gagnon Y (2010) An analysis of feed-in tariff remuneration models: implications for renewable energy investment. Energy Policy 38(2):955–965. doi:10.1016/j.enpol.2009.10.047

Couture TD, Cory K, Kreycik C, Williams E (2010) Policymaker's guide to feed-in tariff policy design. National Renewable Energy Laboratory (NREL), Golden, Colorado, USA

Dong CG (2012) Feed-in tariff vs. renewable portfolio standard: an empirical test of their relative effectiveness in promoting wind capacity development. Energy Policy 42:476–485. doi:10.1016/j.enpol.2011.12.014

Edenhofer O, Pichs-Madruga R, Sokona Y, Seyboth K, Kadner S, Zwickel T, von Stechow C (2011) Renewable energy sources and climate change mitigation: special report of the intergovernmental panel on climate change. Cambridge University Press, Cambridge

Espey S (2001) Renewables portfolio standard: a means for trade with electricity from renewable energy sources? Energy Policy 29(7):557–566. doi:10.1016/s0301-4215(00)00157-9

Fagiani R, Barquin J, Hakvoort R (2013) Risk-based assessment of the cost-efficiency and the effectivity of renewable energy support schemes: certificate markets versus feed-in tariffs. Energy Policy 55:648–661. doi:10.1016/j.enpol.2012.12.066

Galinato GI, Yoder JK (2010) An integrated tax-subsidy policy for carbon emission reduction. Resour Energy Econ 32(3):310–326

Gutermuth PG (1998) Financial measures by the state for the enhanced deployment of renewable energies. Sol Energy 64(1–3):67–78. doi:10.1016/s0038-092x(98)00066-8

Hassett KA, Metcalf GE (1995) Energy tax credits and residential conservation investment: evidence from panel data. J Public Econ 57(2):201–217

IEA (2012) World energy outlook 2012. OECD Publishing, Paris

Jenner S, Groba F, Indvik J (2013) Assessing the strength and effectiveness of renewable electricity feed-in tariffs in European Union countries. Energy Policy 52:385–401. doi:10.1016/j.enpol.2012.09.046

Kahn E (1996) The production tax credit for wind turbine powerplants is an ineffective incentive. Energy Policy 24(5):427–435

Kahn E, Goldman CA (1987) Impact of tax-reform on renewable energy and cogeneration projects. Energy Economics 9(4):215–226. doi:10.1016/0140-9883(87)90029-6

Kalkuhl M, Edenhofer O, Lessmann K (2013) Renewable energy subsidies: second-best policy or fatal aberration for mitigation? Resour Energy Econ 36(3):217–234

Koh KL (2010) Crucial issues in climate change and the Kyoto Protocol: Asia and the world. World Scientific Publishing Company, Singapore

Krajacic G, Duic N, Tsikalakis A, Zoulias M, Caralis G, Panteri E, Carvalho MD (2011) Feed-in tariffs for promotion of energy storage technologies. Energy Policy 39(3):1410–1425. doi:10.1016/j.enpol.2010.12.013

Krey M (2004) Transaction costs of CDM projects in India–an empirical survey. Hamburg, Hamburg Institute of International Economics, HWWA Report 2004:238

Kydes AS (2007) Impacts of a renewable portfolio generation standard on US energy markets. Energy Policy 35(2):809–814. doi:10.1016/j.enpol.2006.03.002

Lauber V (2004) REFIT and RPS: options for a harmonised Community framework. Energy Policy 32(12):1405–1414

Lehmann P (2013) Supplementing an emissions tax by a feed-in tariff for renewable electricity to address learning spillovers. Energy Policy 61:635–641

Lesser JA, Su XJ (2008) Design of an economically efficient feed-in tariff structure for renewable energy development. Energy Policy 36(3):981–990. doi:10.1016/j.enpol.2007.11.007

Levin T, Thomas VM, Lee AJ (2011) State-scale evaluation of renewable electricity policy: the role of renewable electricity credits and carbon taxes. Energy Policy 39(2):950–960

M. J. Beck Consulting L (2009) Renewable Portfolio Standards (RPS). http://mjbeck.emtoolbox.com/?page=Renewable_Portfolio_Standards. Retrieved 26 July 2013

Martinot E, Sawin J (2012) Renewables global status report. Renewables 2012 Global Status Report, REN21. http://www.martinot.info/REN21_GSR2012.pdf

Moon JH, Lee JW, Lee UD (2011) Economic analysis of biomass power generation schemes under renewable energy initiative with Renewable Portfolio Standards (RPS) in Korea. Bioresour Technol 102(20):9550–9557. doi:10.1016/j.biortech.2011.07.041

Nishio K, Asano H (2006) Supply amount and marginal price of renewable electricity under the renewables portfolio standard in Japan. Energy Policy 34(15):2373–2387. doi:10.1016/j.enpol.2005.04.008

Pablo-Romero M, Sánchez-Braza A, Pérez M (2013) Incentives to promote solar thermal energy in Spain. Renew Sust Energ Rev 22:198–208

Proenca S, St Aubyn M (2013) Hybrid modeling to support energy-climate policy: effects of feed-in tariffs to promote renewable energy in Portugal. Energy Economics 38:176–185. doi:10.1016/j.eneco.2013.02.013

Rickerson W, Grace RC (2007) The debate over fixed price incentives for renewable electricity in Europe and the United States: fallout and future directions. A White Paper Prepared for The Heinrich Böll Foundation

Rickerson WH, Sawin JL, Grace RC (2007) If the shoe FITs: using feed-in tariffs to meet US renewable electricity targets. Electr J 20(4):73–86

Rigter J, Vidican G (2010) Cost and optimal feed-in tariff for small scale photovoltaic systems in China. Energy Policy 38(11):6989–7000. doi:10.1016/j.enpol.2010.07.014

Ringel M (2006) Fostering the use of renewable energies in the European Union: the race between feed-in tariffs and green certificates. Renew Energy 31(1):1–17. doi:10.1016/j.renene.2005.03.015

Schallenberg-Rodriguez J, Haas R (2012) Fixed feed-in tariff versus premium: a review of the current Spanish system. Renew Sustain Energy Rev 16(1):293–305. doi:10.1016/j.rser.2011.07.155

Solano-Peralta M, Moner-Girona M, van Sark W, Vallve X (2009) "Tropicalisation" of feed-in tariffs: a custom-made support scheme for hybrid PV/diesel systems in isolated regions. Renew Sustain Energy Rev 13(9):2279–2294. doi:10.1016/j.rser.2009.06.022

Steenblik R, Kim JA (2009) Facilitating Trade in Selected Climate Change Mitigation Technologies in the Energy Supply, Buildings, and Industry Sectors. OECD Library, OECD. http://www.oecd-ilibrary.org/

Tébar Less C, McMillan S (2005) Achieving the successful transfer of environmentally sound technologies. OECD Trade and Environment Working Paper No, 2005:02

Walsh MJ (1989) Energy tax credits and housing improvement. Energy Economics 11(4):275–284

Wand R, Leuthold F (2011) Feed-in tariffs for photovoltaics: learning by doing in Germany? Appl Energy 88(12):4387–4399. doi:10.1016/j.apenergy.2011.05.015

Wiser R (2008) Renewable portfolio standards in the United States-A Status Report with Data Through 2007. LBNL-154E, Lawrence Berkley National Laboratory, Berkley

Wiser R, Porter K, Grace R (2005) Evaluating experience with renewables portfolio standards in the United States. Mitig Adapt Strateg Glob Chang 10(2):237–263

Yin HT, Powers N (2010) Do state renewable portfolio standards promote in-state renewable generation? Energy Policy 38(2):1140–1149. doi:10.1016/j.enpol.2009.10.067

Chapter 6
Market Design for Trading Commoditized Renewable Energy

6.1 Introduction

In the future, energy efficiency will be one of the most important approaches to reducing electricity consumption. For achieving this target, information and communication technology (ICT) will play an important role. In particular, smart grids can achieve energy efficiency by facilitating the interactions between suppliers and customers. Smart grids have been discussed since 2005, when Massoud Amin and Wollenberg (2005) introduced the concept of ICT in electricity networks. Smart grids can help avoid the problems of traditional systems such as transmission loss, inefficiency, unbalanced supply and demand at peak times, and coordinating components of the electricity network. Wang et al. (2011) summarized the current state of research on communication networks of smart grids.

Households that generate electricity using their own renewable energy sources can enter surplus electricity into the electricity network through smart grids. In times of low renewable energy production, they can take electricity from the network. This power exchange between the distributed power sources and the network also requires appropriate management (Mashhour and Moghaddas-Tafreshi 2009). For this, intelligent control systems in households and the smart grid can help. The smart grid can be used to coordinate these activities and enable households to establish a trading system for their surplus electricity. Buchholz and Schluecking (2006) showed different experiences with distributed generation and energy management systems in distribution grids in representative European pilot installations.

Regarding the outlook of the renewable energy market, a marketplace for trading small units of energy is required. By 2030, the US Department of Energy (DOE), through the implementation of its research and development plan, targets to produce 20 % of total electricity generation, an estimated 200 GW, from distributed and renewable energy sources (DOE 2010). Molderink et al. (2009) defined and developed a simulator to analyze the impact of different combinations of microgenerators, energy buffers, appliances, and control algorithms on energy

© Springer Science+Business Media Singapore 2015
A. Heshmati et al., *The Development of Renewable Energy Sources and its Significance for the Environment*, DOI 10.1007/978-981-287-462-7_6

efficiency, both within houses and on a larger scale. Lund et al. (2012) illustrated the reasons for electricity smart grids to be a part of overall smart energy systems and emphasized the inclusion of flexible CHP production in electricity balancing and grid stabilization. In order to facilitate the efficient use of renewable energy sources, power distribution systems at the retail level could change to a market-based operation as it allows aggregation of information about supply and demand in an efficient way. The information exchange can be based in two-way interactions among stakeholders of the electricity marketplace via the smart grid.

Our proposed market provides a place to trade the electricity produced by households within a distribution network. Currently, no marketplace exists for trading these small amounts of electricity. As the marketplace aggregates the demand and supply within the distribution network, it is foreseen to handle the exchange with the transmission network in an efficient way.

The remainder of this chapter gives a brief overview about marketplaces in the electricity industry and highlights the requirements of a marketplace for renewable energy. The requirements for a renewable power marketplace come from the interaction between stakeholders and the market structure.

6.2   Existing Marketplaces

The idea for using a marketplace is not new. It has been employed in wholesale electricity markets and for emission trading. Therefore, we devote this section to discussing marketplaces for trading electricity briefly.

6.2.1   Commercial Marketplaces for Trading Electricity

Three types of marketplaces are available for trading electricity: the spot market, the physical forward market, and the financial futures market. Nord Pool Spot is considered the largest market for electricity worldwide in terms of volume traded and market share. It provides the leading marketplace for buying and selling power in the Nordic and Baltic regions, Germany, and Great Britain. Nord Pool operates three kinds of marketplaces for trading electricity: Elbas (a continuous trading place until 1 h before delivery), Elspot (a continuous trading place until 1 day before delivery), and Eltermin. In Elbas and Elspot, buyers and sellers trade kilowatt-hours. Eltermin, which is the marketplace for trading risk, is divided according to two contract types differing only in the type of the settlement procedure: future and forward (Kristiansen 2007). The marketplace also distinguishes regions in which electricity is traded. The European Energy Exchange (EEX), based in Leipzig, is a leading trading market for energy and energy-related products. The EEX operates a trading platform for electricity, natural gas and CO_2 emission, and coal. The base load financial contracts traded in this market are similar to the Nord Pool contracts (Benth et al. 2008).

6.2.2 Marketplaces for Distributed Electricity Generation

There have been a few studies on marketplaces for understanding the effect of smart grids on prices. Block et al. (2008) introduced a market mechanism that facilitates the efficient matching of electricity and heating demand and supply in microenergy grid environments. Albadi and El-Saadany (2008) presented a summary of demand responses in deregulated electricity markets. They emphasized the effect of demand response on electricity prices by using a simulated case study. Friedman (2002) focused on the technologies required to interconnect DER systems with the grid. Recent increases in electric grid prices coupled with shortages in electricity generation capacity have prompted some industrial and commercial customers to evaluate DER solutions for their energy needs. You et al. (2009a, b, c) proposed a virtual power plant model, which provides individual DER units with access to current electricity markets. They applied this model to microcombined heat and power (μCHP) systems.

6.2.3 Marketplace for Emission Trading

The SO_2 trading system in the USA could be considered an early example of an emission trading system created to reduce the effects of emissions from power plants. The USA has two major emission trading programs: the SO_2 program, which began in the early 1990s, and regional NO_x trading program, which began in the late 1990s (Kruger and Pizer 2004). The EY emission trading system (EU ETS), which started in 2005, is the world's largest emission trading market to date and covers around 50 % of Europe's total CO_2 emissions (Hintermann 2010). As we discussed in Chap. 5, the Kyoto Protocol introduced three mechanisms to reduce GHG emissions: international emissions trading, joint implementation, and a clean development mechanism. The EU ETS is considered an essential part of EU climate change policy in meeting its obligations under the Kyoto Protocol. Hintermann (2010) examined the drivers of allowance prices in the first phase of the EU ETS and found that it was necessary to set an appropriate price to avoid start-up problems. Soleille (2006) argued that ETS itself does not abate emissions. Its efficiency depends on political will, proper design, and implementation. The results obtained by the previous markets in the USA and the EU could be employed in the proposed marketplace in order to take advantage of lessons learned from existing trading programs.

6.3   Stakeholders of the Marketplace

First, we need to recognize the stakeholders of the marketplace for trading renewable energy. Our marketplace includes bulk power generation companies, transmission network operators, distribution network operators, consumers, and small providers.

6.3.1   Bulk Power Generation Companies

Bulk power can be produced by using fossil fuels (e.g., coal and natural gas) and nonfossil fuels (e.g., uranium, water, wind, and sunlight). In addition to this, a virtual power plant (VPP), which combines a number of small DERs with different renewable power types (e.g., wind, solar, and CHP), could be considered to be a bulk power generator. These bulk power providers are connected to the transmission network and thus to the transmission control center through ICT. There is an interactive communication between bulk power providers and the transmission network control center regarding crucial parameters such as capacity, production times, consumption, peak load times, off-peak load times, and related unit costs (Bühler 2010)

6.3.2   Transmission Network Operators

The power generated by bulk power generation companies is transmitted to the distribution network through the transmission network. Only these large power producers and consuming distribution networks are connected to the transmission network operators due to the stability requirements of transmission networks.

6.3.3   Distribution Network Operators

Distribution network operators are responsible for the operation of the electricity network that connects consumers within a small geographic area. For an efficient operation, the distribution network operators would benefit from a smart grid. They could get information about consumption of the connected consumers. Considering that consumers become producers as well, some households may inject their surplus power into the network during some hours and compensate their shortage at other times. In this situation, the distribution network operators demand for transmission network capacity changes (Mashhour and Moghaddas-Tafreshi 2009). This could be a reduced overall consumption with, however, a high variability in the demand over the course of the day (e.g., depending on the weather). The distribution network operator has to cover the missing demand of its consumers by demanding the capacity from the transmission network. Therefore, the role of distribution network operators gains importance if the number of producing consumers increases in the distribution network.

6.3.4 Customers and Small Producers

The consumers not only consume electricity but also generate, store, and sell electricity power. Customers may be households, commercial building owners, or industrial factory owners. These stakeholders play a key role in demand response programs as they can provide substantial demand information to the distribution networks. For this, the consumers need to be connected to the smart grid and control system at their homes. In an advanced state, they can have smart energy-efficient devices and IT architecture at their home. To foster this, there should be some incentives to encourage customers to overcome barriers to acceptance.

6.4 Requirements for a Renewable Power Marketplace

The requirements for designing a successful marketplace for trading renewable power will help establish an environment in which imbalances in demand and supply can be leveled between consumers within distribution networks.

6.4.1 General Economic Requirements

In order to run a successful marketplace for trading renewable power, a few basic economic requirements must be fulfilled. These economic requirements are as follows:

- The perceived risk of accessing resources of many unknown small providers needs to be low.
- The replacement of old devices with new smart grid devices and trading support devices can be performed with low cost. According to the estimation of Faruqui et al. (2010), the cost of building the smart grid for the EU would be 51 billion Euros, and operational saving would be worth 26–41 billion Euros, showing a gap of 10–25 billion Euros between benefits and costs. They argued that smart meters are able to fill this gap as they allow for provision dynamic pricing, which reduces electricity consumption at peak times and lowers the demand for building and operating costly power plants.

- The uncertainty of consumers about the actual cost of resources can be addressed by supplying smart trading software.
- The penetration and liquidity of energy sources is sufficient to bootstrap the marketplace and to attract more market participants.
- There is no barrier for entering the marketplace as a producer and consumer of renewable power.
- The information available to market participants should be equal.
- The pattern of individual demand for renewable power is different for consumers.

If these basic requirements are fulfilled, the market can also have the capacity to provide support and consultancy services, helping customers with power production capacity planning.

6.4.2 Market Structure

Although it is usually difficult for new players to enter an energy market, new players are needed to stimulate the development of the renewable energy industry. Therefore, support policies are needed to provide incentives for suppliers and consumers to use electricity generated by renewable energy sources. Without these support policies, individual demand for electricity can hardly be fulfilled through renewable power sources. With support policies for establishing a market for trading small units of renewable power, the market can develop, show liquidity, and have commodities available.

6.4.3 Electricity Grid Management Requirements

In order to find acceptance for the marketplace for trading renewable power by not only individuals but also transmission providers and bulk power generators, it is crucial that the marketplace is able to meet the following three requirements:

- The electricity supply and demand needs to become predictable.
- The demand for electricity from the transmission networks needs to become more leveled as compared to a situation without a marketplace approach.
- The market participants need a reliable infrastructure for conducting business.

6.4.4 ICT Infrastructure Requirements

To establish this marketplace, ICT infrastructure requires the following technologies to be available:

1. Internet-based networks (i.e., smart grid) and services need to be available for running the smart grid (You 2010). This would interconnect the smart meter and household control software with the marketplace.

2. Because of the intermittent character of wind and solar generators, electricity storage devices (i.e., batteries) are required. Batteries provide dispatch ability, uninterruptible service, and increased efficiency (Tester et al. 2005).
3. Device management needs to be able to connect a newly installed smart meter with the customer's account. It also needs to be able to connect home appliances to the smart meter (Sioshansi 2011).
4. A basic energy management system (EMS) needs to be in place, which is a crucial part of the distributed control (Mashhour and Moghaddas-Tafreshi 2009). The specific instantiation of the EMS is the marketplace itself.
5. Individuals should possess batteries to store renewable energy or the marketplace should provide battery facilities to store renewable energy. It is important that the cost per KWh for storing electricity in a battery goes down.

6.4.5 Regulation Requirements

In order to improve energy efficiency and power generation by renewable energy sources, high-level management is required for the development and implementation of policies and programs (Gellings 2009). There are different kinds of barriers regarding the market (e.g., information transparency, fossil fuel subsidies, financing), technical issues (e.g., lack of skilled workers, knowledge transfer, intellectual property rights), public acceptance (e.g., lack of knowledge, lack of interest, avoidance of comfortable decreases), and the compatibility between current operating systems and new technology implemented in the renewable energy market (e.g., data recording system). These barriers should be removed in order to facilitate the environment for market creation. A reliable political framework ensures both return on investment and continuous research on cost-effective materials, device designing, and improved efficiency (Jäger-Waldau 2006).

As there is no one-type-fits-all regulation, different policies depending on conditions and players' interests are required to achieve this goal. According to the argument of Gunningham et al. (1998), a range of policies is available for environmental protection: command and control regulation, self-regulation, voluntarism, education and information instruments, economic instruments, and free-market environmentalism. Education and information instruments are categorized into education and training, corporate environmental reports, community right-to-know, pollution inventories, product certification, and award schemes. He also categorized economic instruments as property rights, market creation, fiscal instruments, charge systems, financial instruments, liability instruments, performance bonds, deposit refund systems, and the removal of perverse incentives.

As our proposed marketplace includes different stakeholders, we should pay attention to these requirements when we discuss regulation. Furthermore, external factors should be considered in order to maintain the stability of our policy and avoid causing uncertainty among the market participants regarding policies.

In order to establish a marketplace for trading renewable power successfully, we recommend fulfilling a combination of the abovementioned instruments as a regulatory framework:

- Command and control: These regulations apply specific standards to energy consumption in industries, public sectors, commercial buildings, and households. This regulation could be used in combination with awards and penalty regulations.
- Self-regulation: The government sets a specific standard for industries, and every industry self-regulates to achieve this standard. This kind of regulation is also seen at the international level. For example, the OECD sets a target for members to reduce CO_2 emissions, and each member self-regulates to achieve organizational targets.
- Education and information: Education and information form a crucial parameter in developing the capacity of renewable energy usage in industry and the community. Public acceptance is one of the most important components of market development. Environmental information and training programs presented by the government could be considered as a supplementary instrument to other forms of regulation (Gunningham et al. 1998).
- Economic instruments: A wide range of economic instruments exists to encourage private companies and consumers to use renewable energy sources. These include feed-in tariffs, reduction of fossil fuel subsidies, CO_2 emission trading, renewable fuel standards or targets, green certificate trading, emission and energy taxes, residential and commercial tax credits for renewable energy usage, and the Kyoto Protocol (Hofman and Huisman 2012). It is recommended that marketplace creation be stimulated by feed-in tariffs.
- Stability: Any regulation set by policymakers should be stable. In Chap. 4, we discussed that in addition to grid modernization efforts to enhance energy efficiency and adapt to the current policy, a regulatory framework and market environment are crucial in supporting new technology investments. However, suppliers and consumers will not be attracted if they cannot trust the policies due to uncertainty. For example, the financial crisis has forced some European governments using feed-in tariffs to cut their subsidies (Hofman and Huisman 2012). Public views and investor decision-making are negatively affected by this kind of behavior. Therefore, environmental policies and all related supportive instruments, especially economic ones, should be selected properly and considered high priority in governmental planning.

6.4.6   Ease-of-Use Requirements

The marketplace needs to set up an environment that is trusted by market participants. In particular, the following five requirements are recommended:

- In order to guarantee privacy, energy providers do not give customer information to third parties.

- The marketplace provides access to power resources in a transparent and simple way.
- A user-friendly interface to the marketplace exists enabling market participants to trade and communicate easily.
- The technology for commoditizing energy resources is easily usable by all people, even those who are not highly educated. Many people in urban areas, especially in suburban areas, are neither familiar with smart grid technology nor marketplaces.
- Sufficient information about electricity price, electricity consumption, and electricity production is available to all customers.

6.5 Market Mechanism for Trading Renewable Power

The task of the market mechanism is to match offers to sell excess renewable power with orders for renewable power that could not be covered with in-house renewable power sources. Different market mechanisms can be used for trading. Continuous double auction (CDA) type, which is used widely in stock exchanges and commodity markets, is well suited as the CDA market mechanism matches buyers and sellers of a particular good continuously. A match occurs if the sell price is lower than or equal to the buy price. The trading of renewable power differs from other wholesale electricity markets due to the small amount of power that can be traded at the retail market.

The definition of the unit of trade and the definition of bids and asks is based on Altmann et al. (2008) and Altmann et al. (2010).

6.5.1 Unit of Trade

The unit of trade is defined through the following parameters:

- Start time: It is the time at which the resource is available for the buyer or the time that the resource is required by the buyer.
- Unit duration: This is the standard length of time that the resource will be available to the buyer or the shortest period that the resource will be required by the buyer. The unit duration is set according to the acceptance of users within the marketplace. The unit of trade is 1 h in the wholesale power market. At the retail market, because of the short duration of activities performed at home, we use smaller unit durations. A unit duration of 6 min is commonly used as the minimum duration of consumption of appliances.
- Unit volume: This is set to 20 W according to the consumption of a low-end appliance at home.

Although the unit of trade specified is very small, it is expected not to cause computational problems as any battery use can reduce the trading significantly. In addition to this, the bids and asks that are described in the following section allow reducing the computational complexity even further.

6.5.2 Bids and Asks

Based on the unit of trade, we define bids and asks. Ask is submitted by a producer who owns extra electricity and wants to supply it to the market. The bid is submitted by a customer who needs extra electricity at a low price. The bids and asks comprise the following parameters:

- Price: This parameter defines the minimum price that a seller is willing to accept or the maximum price a buyer is willing to pay for a unit of trade. Any fees that the marketplace charges are implicitly included as a cost factor in the prices.
- Volume of unit-of-trades: This parameter defines the number of units of trade that is to be traded. As the actual unit of trade is quite small, each trader may want to exchange larger chunks of electricity.
- Duration: This parameter defines the overall duration of the volume to be traded. It is a multiple of the basic duration of a unit of trade. As the actual unit of trade is quite small, a trader can specify a longer time period than the basic unit of the unit of trade.

6.6 Performance Evaluation

Although the investments in many projects involving energy-efficient technology show good economic results, the percentage of their successful implementation is less than expected due to barriers that discourage decision-makers such as households and firms (IEA 2012). The lack of a measurement of energy efficiency has led to the effect that opportunities are not visible. Therefore, a definition of some performance indices becomes necessary to show households, firms, and policymakers the economic and environmental benefits of a market-based approach to renewable power trading at the distribution network level.

Some indices are used in current markets. For example, Energy Exchange Austria (EXAA) is a European energy exchange headquartered in Vienna covering energy trading in Austria and Germany. Table 6.1 shows some selected indices used in EXAA, which are also applicable to the market-based exchange of renewable power at the distribution network–level grid market.

Table 6.1 Key performance ratios for EXAA

Key performance ratios		2010	2011
Sales revenue in Euro		2,030,159	2,324,493
Spot market electric power	Trading volume in GWh	6,410	7,558
	Clearing volume in Euro	292,146,570	390,236,567
	Number of trading members	90	71
Spot market CO_2 allowance	Trading volume in t	88,401	19,179
	Trading volume in Euro	1,260,481	269,072
Market share in \% of Austrian consumption		10.7	13.1

Source: *Annual Report 2011, EXAA*

In addition to the indices used for the energy exchange center, other indices such as storage capacity, storage inflow, storage outflow, generation capacity, installed generators, and load shift numbers could be calculated to analyze market performance

References

Albadi MH, El-Saadany E (2008) A summary of demand response in electricity markets. Electr Power Syst Res 78(11):1989–1996

Altmann J, Courcoubetis C, Stamoulis G, Dramitinos M, Rayna T, Risch M, Bannink C (2008) GridEcon: a market place for computing resources. Grid Econ Bus Model 185–196

Altmann J, Courcoubetis C, Risch M (2010) A marketplace and its market mechanism for trading commoditized computing resources. Ann Telecommun 65(11–12):653–667

Benth FE, Benth JS, Koekebakker S (2008) Stochastic modelling of electricity and related markets, vol 11, Advanced series on statistical science and applied probability. World Scientific Publishing Company Incorporated, Hackensack

Block C, Neumann D, Weinhardt C (2008) A market mechanism for energy allocation in micro-chp grids. Paper presented at the Hawaii International Conference on System Sciences, Proceedings of the 41st Annual, Universität Karlsruhe, Germany

Buchholz B, Schluecking U (2006) Energy management in distribution grids European cases. Paper presented at the Power Engineering Society General Meeting, 2006. IEEE

Bühler R (2010) Integration of renewable energy sources using microgrids, virtual power plants and the energy hub approach. Swiss Federal Institute of Technology, Zurich

DOE U (2010) Smart grid research & development: multi-year program plan (MYPP) 2010–2014, U.S. Department of Energy, Office of Electricity Delivery & Energy Reliability

Faruqui A, Harris D, Hledik R (2010) Unlocking the € 53 billion savings from smart meters in the EU: how increasing the adoption of dynamic tariffs could make or break the EU's smart grid investment. Energy Policy 38(10):6222–6231

Friedman NR (2002) Distributed energy resources interconnection systems: technology review and research needs. National Renewable Energy Laboratory, NREL, Golden, Colorado

Gellings CW (2009) The smart grid: enabling energy efficiency and demand response. Fairmont Press, Lilburn

Gunningham N, Grabosky P, Sinclair D (1998) Smart regulation: designing environmental policy. Oxford Clarendon Press, Oxford

Hintermann B (2010) Allowance price drivers in the first phase of the EU ETS. J Environ Econ Manag 59(1):43–56

Hofman DM, Huisman R (2012) Did the financial crisis lead to changes in private equity investor preferences regarding renewable energy and climate policies? Energy Policy 47:111–116

IEA (2012) Medium-term renewable energy market report 2012. OECD Publishing, Paris

Jäger-Waldau A (2006) European Photovoltaics in world wide comparison. J Non-Cryst Solids 352(9):1922–1927

Kristiansen T (2007) Pricing of monthly forward contracts in the Nord Pool market. Energy Policy 35(1):307–316

Kruger J, Pizer WA (2004) The EU emissions trading directive: opportunities and potential pitfalls, Resources for the future, RFF Discussion Paper 2004:24

Lund H, Andersen AN, Østergaard PA, Mathiesen BV, Connolly D (2012) From electricity smart grids to smart energy systems–a market operation based approach and understanding. Energy 42(1):96–102

Mashhour E, Moghaddas-Tafreshi S (2009) A review on operation of micro grids and Virtual Power Plants in the power markets. Paper presented at the Adaptive Science & Technology, 2009. ICAST 2009. 2nd International Conference, Accra

Massoud Amin S, Wollenberg BF (2005) Toward a smart grid: power delivery for the 21st century. Power Energy Magazine, IEEE 3(5):34–41

Molderink A, Bosman MGC, Bakker V, Hurink JL, Smit GJM (2009) Simulating the effect on the energy efficiency of smart grid technologies. Paper presented at the Winter Simulation Conference, University of Twente, Netherland

Sioshansi FP (2011) Smart grid: integrating renewable, distributed & efficient energy. Academic, Burlington

Soleille S (2006) Greenhouse gas emission trading schemes: a new tool for the environmental regulator's kit. Energy Policy 34(13):1473–1477

Tester JW, Drake EM, Driscoll MJ, Golay MW, Peters WA (2005) Sustainable energy: choosing among options. The MIT Press, Cambridge, MA

Wang W, Xu Y, Khanna M (2011) A survey on the communication architectures in smart grid. Comput Netw 55(15):3604–3629

You S (2010) Developing virtual power plant for optimized distributed energy resources operation and integration. PhD thesis, Technical University of Denmark, 2011. [2] SENERTEC, "Technical Documentation CHP unit-DACHS HKA, SENERTEC", vol. Art. Nr. 00/4798

You S, Træholt C, Poulsen B (2009a) Generic virtual power plants: management of distributed energy resources under liberalized electricity market. Paper presented at the Advances in Power System Control, Operation and Management (APSCOM 2009), 8th International Conference on Technical University of Denmark, Kgs. Lyngby, Denmark

You S, Træholt C, Poulsen B (2009b) A market-based virtual power plant. Paper presented at the Clean Electrical Power, 2009 International Conference on Technical University of Denmark, Kgs. Lyngby, Denmark

You S, Traholt C, Poulsen B (2009c) A study on electricity export capability of the μCHP system with spot price. Paper presented at the Power & Energy Society General Meeting, 2009. PES'09. IEEE

Chapter 7
Impact of Renewable Energy Development on Carbon Dioxide Emission Reduction

7.1 Introduction

During the last three decades, two different approaches have been applied in the research of natural resources. The first approach considers the effect of natural resources on economic growth. As previously discussed in Chap. 2, many researchers have studied the relationship between energy consumption and economic growth. Early studies were published in the 1970s, including Allen et al. (1976), Hitch (1978), and Kraft and Kraft (1978). This relationship has been studied in both individual and groups of countries. Akarca and Long (1980), Yu and Hwang (1984), Cheng (1995), and Stern (2000) applied this methodology in the USA. Wolde-Rufael (2005) employed the causal relationship methodology in a study of 19 African countries. Lee and Chang (2008) tested causality in 16 Asian countries, and Huang et al. (2008) tested causality in 82 countries. The effect of energy consumption on economic growth could differ substantially in developed countries. Narayan and Prasad (2008) found different causality effects in 30 OECD countries. This means that conservative energy policy could affect individual countries differently.

The second approach takes into account the environmental effects of economic growth. Following the empirical study of Grossman and Krueger (1991), many scholars analyzed the relation between economic growth and environmental pollution. Coondoo and Dinda (2002) studied the relationship of income to CO_2 emissions based on a Granger causality test of cross-country panel data. Zhang and Cheng (2009) investigated Granger causality between economic growth, energy consumption, and CO_2 emissions in China. Soytas and Sari (2009) examined causality relationships between these variables in Turkey. In the most recent study, Choi et al. (2010) used the environmental Kuznets curve (EKC) to examine the relationship of CO_2 with economic growth and openness.

In the present study, we investigate EKC as well. However, we consider a number of variables, namely, the share of renewable energy sources in total power

© Springer Science+Business Media Singapore 2015   119
A. Heshmati et al., *The Development of Renewable Energy Sources and its Significance for the Environment*, DOI 10.1007/978-981-287-462-7_7

generated, the number of patents per million inhabitants for energy applications adopted to mitigate climate change, the number of ICT patents as a proxy variable for technological innovation, environmental tax per capita as a proxy variable for market regulation, and trends within the data set. We examine the effects of these variables on CO_2 emissions per capita.

Our contribution is to investigate the effectiveness of power generated through renewable energy sources, technological innovation, and market regulations on the mitigation of climate change. We also calculate the elasticity of CO_2 emissions per capita for each parameter. The results obtained by our model could be used by policymakers to evaluate the effectiveness of different policy tools and the effects of interactions between these policies.

7.2 Methodology and Analytic Framework

In this section, we explain the methodology used to build the model, we define the model specifications, and we state the hypothesis of our model.

7.2.1 Methodology

A cross-country panel data model has been applied to EU-15 countries. Hsiao (2003) and Klevmarken (1989) mentioned various advantages for using panel data, such as controlling for individual heterogeneity, more variability, less colinearity among the variables, more degrees of freedom, and more efficiency (Baltagi 2008). The countries differ in terms of economic structure, technology, and policy. If this heterogeneity is not included in the model, serious misspecification could result. Moreover, the probability of colinearity is high in time series studies, but it is less likely with a panel across countries because the cross-country dimension adds variability. Additionally, informative data could lead to increased reliability of estimators. Several different linear models can be applied to panel data. The individual-specific-effects model for the dependent variable, y_{it}, specifies that

$$y_{it} = X_{it}\beta + \alpha_i + \gamma_t + v_{it} \tag{7.1}$$

where α_i and γ_t are error components (or random effects) specific to units i and time periods t. They are individual and time-specific effects. The composite error term $u_{it} = \alpha_i + \gamma_t + v_{it}$ is generally not independent and identically distributed (i.i.d.), but its variance–covariance matrix (Ω) could be estimated. y_{it} is the NT-by-1 vector of a dependent variable, X_{it} is the NT-by-k matrix of independent variables, and β is the k-by-1 vector of unknown coefficients that is estimated. Two different assumptions may be applied for α_i and γ_t in a large proportion of empirical applications: fixed-effect and random-effect models (Johnston and DiNardo 2007). In the fixed-effect

(FE) model, α_i and γ_t can be correlated with the independent variables X, but it is assumed that X_{it} is uncorrelated with the random error term v_{it}. The attraction of the FE model is that it follows a consistent estimator (convergence in probability) (Cameron and Trivedi 2009). On the other hand, the random effects (RE) model assumes that α_i and γ_t are random and not correlated with the independent variables.

Based on the Gauss–Markov theorem, if the errors have expectation zero, are uncorrelated, and have equal variance, the ordinary least square (OLS) estimator have minimum variance in the class of linear unbiased estimators. It is called the best linear unbiased estimator, or BLUE (Johnston and DiNardo 2007). Therefore, we have these assumptions:

$$E\left(u\middle|X\right) = 0 \qquad (7.2)$$

$$E\left(uu'\middle|X\right) = \sigma^2\Omega \qquad (7.3)$$

In Eq. (7.2), the disturbances have conditional zero mean. In Eq. (7.3), $\Omega = I_N$ is an N-by-N identity matrix. It means that conditional on the X, the disturbances are independent and identically distributed with conditional variance σ^2 (Jackman 2004). Then, the ordinary least estimator $\widehat{\beta}_{OLS} = (X'X)X'y$ with variance–covariance matrix $V\left(\widehat{\beta}_{OLS}\right) = \sigma^2(X'X)^{-1}$ is the best linear unbiased estimator (BLUE). If the assumption in Eq. (7.3) fails to hold, the mentioned estimator for the parameters is unbiased, but it is not BLUE. According to Johnston and DiNardo (2007), Eq. (7.3) states that disturbances have homoscedasticity and are pairwise uncorrelated. However, this condition is rarely satisfied in practice, so it is important to develop feasible generalized least square (FGLS) estimators, where unknown parameters are substituted by consistent estimates (Johnston and DiNardo 2007).

It should be noted that when Eq. (7.3) holds, $\Omega = I_N$ and $\widehat{\beta}_{GLS} = \widehat{\beta}_{OLS}$. Considering that we usually do not have knowledge about Ω, $\widehat{\beta}_{GLS}$ is nonoperational, and we have to utilize an FGLS estimator. FGLS estimators are calculated in three steps (Jackman 2004):

1. OLS analysis to yield estimated residuals $\hat{u}$
2. Analysis of the $\hat{u}$ to form an estimate of Ω
3. Calculation of the FGLS estimator as $\widehat{\beta}_{FGLS} = \left(X'\widehat{\Omega}^{-1}X\right)^{-1}X'\widehat{\Omega}^{-1}y$

FGLS is the most commonly used estimator in dealing with residual autocorrelation and heteroscedasticity. Cochrane and Orcutt (1949) and Prais and Winsten (1954) established procedures for AR (1) disturbances yielding FGLS estimators. Applying FGLS to deal with heteroscedasticity has been mentioned in different econometric texts. For example, Judge et al. (1988: 128–145) and Amemiya (1985: 198–207) provided different rigorous treatments of FGLS estimators in this regard (Jackman 2004).

7.2.2 Model Specification

We specify and estimate our model based on the EKC. The Kuznets curve was introduced for the first time by Simon Kuznets in 1955 to show the relationship between inequality in the distribution of income and levels of income (Kuznets 1955). However, in the 1990s, the curve became an engine for studying the relationship between emissions and economic growth. The standard form of this function is defined as follows (Grossman and Krueger 1991):

$$\ln (E/P)_{it} = \alpha_i + \beta_1 \ln (GDP/P)_{it} + \beta_2 \Big(\ln (GDP/P)^2\Big)_{it} + \varepsilon_{it} \qquad (7.4)$$

where E is urban air pollution, P is population, GDP is gross domestic product, and ln indicates natural logarithms. Several basic models have been estimated without additional independent variables (Grossman and Krueger 1991; Shafik and Bandyopadhyay 1992; Selden and Song 1994). Many researchers have studied this model, using additional explanatory variables to evaluate the environmental effects of different factors. Panayotou (1993) examined the hypothesis of deforestation (DEF) as a function of income per capita (INC) and population (POP) density as follows:

$$\ln (DEF) = \alpha_1 \ln INC + \alpha_2 \ln POP + \frac{1}{2} \alpha_{11} (\ln INC)^2 + \frac{1}{2} \alpha_{22} (\ln POP)^2$$
$$+ \alpha_{12} (\ln INC) (\ln POP) + \varepsilon_{it} \qquad (7.5)$$

Panayotou applied a translog formulation to allow for the effects of interaction between explanatory variables and the evaluation of elasticities. The interaction terms' coefficient signs indicate substitution ($-$) and complementarity ($+$) relationships between the explanatory variables in their effects on the DEF. The square terms also provide information about nonlinear relationships between the explanatory variables and the dependent variable.

Some researchers investigated the effects on environmental quality of literacy, political rights, civil liberties (Torras and Boyce 1998), output structure (Panayotou 1997), and trade (Suri and Chapman 1998). Choi et al. (2010) employed trade dependence, fossil consumption per capita, share of renewable energy, and time trend in an attempt to broaden the concept of EKC and evaluate the impact of these parameters on CO_2 emissions. Magnani (2000) used R&D expenditure as a proxy to measure the impact of environmental protection on pollution emissions and to analyze the EKC.

In this study, we evaluate the impact of renewable energy deployment, technological innovation, and market regulation on CO_2 emission reduction. Similar to Panayotou (1993), our model is formulated in a translog function form in order to investigate the interaction effects between variables and calculate elasticities. Our model also permits us to check the shape of the relationship between dependent variables and explanatory variables:

$$\ln Y_{it} = \alpha_0 + \sum_j \beta_j \ln X_{jit} + \frac{1}{2} \sum_j \beta_{jj} \ln X_{jit}^2 + \sum_i \sum_j \beta_{jk} \ln X_{jit} \ln X_{kit}$$

$$+ \alpha_t t + \alpha_{tt} t^2 + \sum_j \beta_{jt} \ln X_{jit} t + \varepsilon_{it}, \tag{7.6}$$

where the variable Y represents the dependent variables defined as CO_2 emissions per capita (CO_2/P), the matrix X represents a set of independent variables including gross domestic product per capita (GDP/P), share of electricity generated by renewable energy sources in total power generation (Ren/TPG), number of energy-related patents per million inhabitants (Pateng/P), number of ICT patents per inhabitant (Patict/P), and environmental tax per capita (Evt/P), and t is a time trend (Trd) representing the rate of technical change or shift in the CO_2 function over time as a result of technological progress or change. It captures percentage reduction in CO_2 per year.

7.2.3 Analytic Framework for Variable Selection

Considering the long lifetime of CO_2 in the atmosphere, stabilizing the concentration of GHGs at any level depends on large reductions of worldwide CO_2 emissions from current levels (IEA 2012). Therefore, CO_2 emission reduction could be used as an index to evaluate climate change mitigation. In our model, CO_2 emission is considered a dependent variable, and it is defined as total CO_2 emissions caused by the consumption of energy. The explanatory variables are defined as follows:

Gross domestic product (GDP): As discussed in Chap. 2, there is extensive literature on the relationship between economic growth and energy consumption. Consistent with the literature about the relationship between GDP and environmental pollution described in Sect. 7.2.2, we have considered GDP as an explanatory variable in our model.

Renewable energy sources: We have used the share of electricity produced by renewable energy sources (variable Ren) in total power generation (variable TPG) in our model, which is in line with the literature on the effectiveness of renewable energy sources in climate change mitigation (Sinha 1993; Frankl et al. 1997; Schleisner 2000; Lehner et al. 2005; Benitez et al. 2008; Saner et al. 2010) and the contribution of renewable energy in power generation as an effective parameter of CO_2 emissions (Choi et al. 2010). According to the US Energy Information Administration, renewable energy sources include biomass, hydro, geothermal, solar, wind, ocean thermal, wave action, and tidal action (EIA 2013).

Energy patent applications: Innovation has responded to emission abatement expenditure over time (Lanjouw and Mody 1996; Popp 2003). Innovation could be done through building energy efficiency (Li and Colombier 2009) or technological change in renewable energy capacity (Popp et al. 2011). Therefore, considering that technological innovation plays an important role in mitigating the effects of environmental problems, we added the number of energy technology patents (variable Pateng) to our model. These are energy technology patent applications to the European Patent Office (EPO) for mitigation or adaptation to climate change, including capture, storage, sequestration, or disposal of GHGs, and the reduction of GHG emissions related to energy generation, transmission, or distribution.

ICT patent applications: Energy efficiency is a solution used to reduce CO_2 emissions. As discussed in Chap. 3, different technologies could be used for this purpose, such as electric vehicles (Ford 1995; Kempton and Letendre 1997; Kempton and Tomić 2005), virtual power plants (Pudjianto et al. 2007; Ruiz et al. 2009; Jansen et al. 2010), and smart meters (Hartway et al. 1999; Faruqui et al. 2007; Depuru et al. 2011). Considering that patent applications for ICT play a crucial role in facilitating these technologies, we employed ICT patent (variable Patict) applications to the EPO as another kind of technological innovation in our model.

Environmental tax: Environmental tax (Evt) is used in our model as a proxy for market regulation in order to evaluate its impact on CO_2 emission reduction. Nordic countries, such as Finland and Sweden, are in the forefront of taxing fuels because of environmental damage (Bhattacharyya 2011). In line with the literature and based on the findings (Alfsen et al. 1995; Brännlund and Nordström 2004; Galinato and Yoder 2010) regarding the effectiveness of carbon tax policy on CO_2 emission reduction, we used environmental tax as an explanatory variable in our model. Total environmental taxes are for energy products (including fuel for transport), transport (excluding fuel for transport), pollution, and natural resources (excluding oil and gas). Based on Eurostat, CO_2 taxes are included under energy taxes instead of pollution taxes. The reason is that in many cases, CO_2 taxes are levied on the same tax base as energy is. Therefore, considering CO_2 taxes as a pollution tax instead of an energy tax would distort international comparisons.

Trend: Time trend represents the possibility that technological change causes effects or shifts in the environment function over time (Shafik and Bandyopadhyay 1992; Cole et al. 1997; Luzzati and Orsini 2009). It is expected that the CO_2 function will shift downward, suggesting progress or reduction in CO_2 over time for given energy use and GDP production. The trend is included in the specifications of the model to allow for the possibility of technological change or shift in the CO_2 model over time. The time trend is generally taken into account as a proxy for technological effects (Shafik and Bandyopadhyay 1992). The inclusion of trend squared allows testing for the nonlinearity of the shift or changes in the CO_2 function, which is more realistic considering the amount of production and energy use. Technological progress is expected to impact CO_2 negatively. The coefficient shows the changes in the dependent variable holding constant the influence of the other explanatory variables. The coefficient could also reflect some unmeasured factors in trend, such as increasing productivity and efficiency in production, which may affect CO_2 emissions over time.

Fossil fuel energy is a major source of CO_2 emissions. However, this variable is not used to avoid the colinearity problem. Considering that we have employed

GDP in our model, the related effects could be captured through this variable. Fossil fuel is the major source of primary energy consumption in the EU-15. According to *British Petroleum Statistical Review of World Energy*, more than 90 % of the primary energy used in the EU-15 comes from nonrenewable energy sources (BP 2012). If fossil fuel consumption is used as an independent variable in our model, we may have colinearity between GDP and fossil fuel consumption.

By applying the translog formulation, we can examine the interactions among these variables and their potential effects on policymaking. For example, an increase in environmental compliance cost could lead to increases in the patenting of new environmental technologies (Jaffe and Palmer 1997). Therefore, they could have an interaction effect on CO_2 emission reduction. According to Stern (1998), the use of resources implies the production of waste. Therefore, the regressions that consider indicators that are allowed to become zero or negative are not estimated appropriately. We applied restriction by using a logarithmic dependent variable in order to comply with Stern's comment.

7.2.4 Hypotheses

We have examined the effect of renewable energy development (Chap. 2), technological innovation (Chap. 3), and market regulation applied by governments (Chap. 5). We, therefore, define the following three hypotheses:

1. The power generated by renewable energy sources in the EU-15 has been able to affect CO_2 emissions by displacing traditional capacity fueled by fossil fuels. We also expect negative elasticity for renewable energy sources regarding CO_2 emissions.
2. Technological advances are able to decrease the costs of renewable energy technologies and energy efficiencies, thereby saving energy and reducing CO_2 emissions. Therefore, we expect a negative relation between technological innovation and CO_2 emissions. Furthermore, we expect negative elasticity for technological innovation.
3. Environmental tax applied by governments has a direct negative relation with CO_2 emissions. The size of this parameter could indicate its importance, compared with renewable energy development and technological innovation. We also expect a negative elasticity for environmental tax.

It should be noted that the way the model is specified, it allows to separate the technological effects into two parts. The first part of technology which is observable and related to patents can be separated from the unobservable part captured by the time trend.

Elasticity is the measurement of how changing one variable affects CO_2 emissions per capita. This measurement enables policymakers to know the effectiveness of each policy, which helps them to make appropriate decisions to achieve targets set for CO_2 emission reduction.

7.2.5   Top–Down Versus Bottom–Up Approach

Technology-oriented studies use bottom–up engineering models, which are based on the integration of data on the cost and performance of technology. Economic studies employ top–down models to analyze aggregate behavior according to economic indices and elasticities, and they focus on applying carbon tax to limit emissions (Grubb et al. 1993). As discussed in Chap. 3, most researchers have used the bottom–up approach to show the potential of carbon saving obtained by different renewable energy technologies. These studies concluded that CO_2 emission reduction could be achieved by carbon saving. Previous research noted that differences between top–down and bottom–up approaches are less theoretical and relate to the level of aggregation to assumptions (Böhringer 1998).

The bottom–up approach is applied by researchers to show the lower cost of CO_2 emission reduction compared to top–down estimates. Williges et al. (2010) examined the cost of a European feed-in tariff for the large-scale development of concentrated solar power. Their results were significantly lower than most current top–down estimates of achieving GHG-stabilizing scenarios. However, the bottom–up approach is applied in pure engineering, and it is not an appropriate model for policymaking and market behavior. Some important parameters such as hidden costs, the cost of implementation measures, market imperfection and other economic barriers, and macroeconomic relationships are not included in these models (Grubb et al. 1993).

Because this study evaluates the effects of renewable energy development, technological innovation, and market regulation on CO_2 emission reduction in addition to the effects of their interactions, the top–down approach is appropriate. It should be noted that the bottom–up approach could be applied in the selection of competitive renewable energy technology when a proper policy has been made. In other words, a combination of the top–down and the bottom–up approach is required to achieve targets for CO_2 emission reduction. Governmental commitment to fostering renewable energy sources, with a combination of bottom–up and top–down approaches, was the key to initial success in Denmark (Lipp 2007).

7.3   Model Estimation

In this section, we compare our proposed model with two other forms of function formulation and evaluate it by testing the functional form, model specification, and share of significance of the parameters. We then analyze the results and parameters. Finally, we present the conclusion and policy recommendations.

7.3.1 Data

In this research, we investigate the effects of renewable energy development, technological innovation, and market regulation on CO_2 emission reduction in the EU-15 countries, which include Austria, Belgium, Germany, Denmark, Spain, Finland, France, Greece, Ireland, Italy, Luxembourg, the Netherlands, Portugal, Sweden, and the UK. We selected these countries because they are at the forefront of renewable energy development and applied market regulation to mitigate climate change. In 1990, Finland was the first country to apply a carbon tax, followed by Norway, Sweden, and Denmark in 1992 (Bhattacharyya 2011). Germany's renewable sector is considered the most innovative and successful in the world. Since starting negotiations on climate change in 1991, the EU has provided leadership in global climate policy (Oberthür and Roche Kelly 2008).

These countries are studied during the period from 1995 to 2010. Because the impact of the financial crisis began in 2008, it has been included in the model. We used different sources to obtain information for our model. The data on CO_2 emissions, total power generation, and electricity produced by renewable energy sources was obtained from the US Energy Information Administration (EIA 2013) database. The information related to environmental tax and the number of patent applications in energy technology and ICT was derived from the European Commission database (Eurostat). Population sizes were extracted from the World Bank database. Table 7.1 shows the summary, definition, and source of data for all variables, including CO_2/P (cdecap), GDP/P (gdpcap), Ren/TPG (regenp), Pateng/P (ptgcap), Patict/P (pticap), Evt/P (evtcap), and trend.

Table 7.1 Summary of the independent variables, their definitions, and data sources

Variable	Definition	Source
CDECAP	CO_2 emissions per capita made by fossil fuels burning and the cement manufacturers. CO_2 made by consumption of solid, liquid, and gas fuels and gas flaring are included (metric ton per capita).	WB
REGENP	Share of electricity produced by renewable energy sources (including biomass, hydro, geothermal, solar, wind, ocean thermal, wave action, and tidal action) in total power generation (percent)	EIA
GDPCAP	Gross domestic product per capita based on 2005 constant dollar (1,000 USD per capita)	WB
PTGCAP	Number of patents in energy technologies or applications for mitigation or adaptation against climate change per million inhabitants	Eurostat
PTICAP	Number of patents in information and communication technology (ICT) per million inhabitants	Eurostat
EVTCAP	Total environmental tax per capita based on 2005 constant dollar, including energy products, transport, and pollution (1,000 USD per capita).	Eurostat
TREND	Time trend	

Table 7.2 Correlation matrix between explanatory variables (p-values in parenthesis)

	CDECAP	GDPCAP	REGENP	PTGCAP	PTICAP	EVTCAP	TREND
CDECAP	1.0000						
GDPCAP	0.7085	1.0000					
	(0.0000)						
REGENP	0.0.640	0.1642	1.0000				
	(0.3233)	(0.0109)					
PTGCAP	0.1441	0.4685	0.0715	1.0000			
	(0.0256)	(0.0000)	(0.2699)				
PTICAP	0.0516	0.0843	−0.1138	0.1087	1.0000		
	(0.4260)	(0.1933)	(0.0785)	(0.0931)			
EVTCAP	0.1798	0.2749	−0.0876	0.3829	0.0934	1.0000	
	(0.0052)	(0.0000)	(0.1763)	(0.0000)	(0.1492)		
TREND	0.0017	0.3479	−0.0661	0.3472	0.0283	0.1517	1.0000
	(0.9796)	(0.0000)	(0.3079)	(0.0000)	(0.6623)	(0.0187)	

Because the variables are measured in different units, the continuous variables are transformed to logarithmic form. Therefore, the coefficients are interpreted on the basis of the percentage of change. Regarding the unit measured for the variables, the same unit was used for the explanatory variables. GDP per capita and environmental tax per capita were measured by a constant US dollar (2005). They were normalized by the deflator index. The measurement of the energy patents and ICT technologies is based on the number of patents per million inhabitants.

Table 7.2 displays the correlation matrix or the unconditional pairwise correlation relationship between the variables presented in Table 7.1. We constructed this table in order to check the existence of correlations between variables. If a variable shows a high correlation (a correlation coefficient larger than 0.50) and is significant for another variable, it should be omitted in order to avoid colinearity. As the correlation matrix shows, no significant large correlation was found between variables indicating no serious colinearity problem.

The second line of each variable in parenthesis presents the p-values of correlation coefficients and significant coefficients at the level of 5 %. The coefficient for some correlations is not significant. The sign for some coefficients such as CO_2 emissions per capita–gross domestic production per capita (cde-gdpcap), environmental tax per capita–number of energy patents per million inhabitants (evtcap-ptgcap), trend-gdpcap, and environmental tax per capita–gross domestic production per capita (evtcap-gdpcap) are as expected. However, others, such as CO_2 emissions–energy patent (cde-ptg) and CO_2 emission per capita–environmental tax per capita (cdecap-evtcap), are not as expected. We should consider that the correlation matrix presents the unconditional relation between variables. This means that the sign and significance situation could change when they are studied with the other variables (conditional). Moreover, the correlation matrix examines the

Table 7.3 Fisher-type unit root test result

Variable	Inverse chi-squared	Inverse normal	Inverse logit t(79)	Modified inv. Chi-squared
LCDECAP	102.4375	−6.9248	−7.1988	9.3516
LGDPCAP	121.3020	−7.2424	−8.4251	11.7870
LPTGCAP	154.5430	−9.4482	−11.0220	16.0784
LPTICAP	108.2611	−6.3102	−7.1981	10.1035
LEVTCAP	101.8071	−6.5370	−7.0435	9.2703
LREGENP	79.0775	−5.0988	−5.2345	6.3359
LGDPCAP2	116.7426	−7.0275	−8.0665	11.1984
LREGENP2	60.1541	−3.3450	−3.3837	3.8929
LPTGCAP2	130.8208	−8.4524	−9.3066	13.0159
LPTICAP2	103.8467	−5.6432	−6.5478	9.5336
LEVTCAP2	63.0009	−4.0403	−4.0083	4.2604
GDPRENP	87.7950	−5.7567	−5.9721	7.4613
GDPPTIP	109.2815	−6.1472	−7.0694	10.2352
GDPENGP	154.4527	−9.4641	−11.0214	16.0668
GDPEVTP	93.2944	−5.7185	−6.2946	8.1713
RENPTGP	152.1289	−9.3607	−10.8274	15.7668
RENPTIP	63.3171	−3.4676	−3.5652	4.3012
RENEVTP	74.5212	−4.5670	−4.8312	5.7477
PTGPTIP	147.9147	−9.2704	−10.5574	15.2227
PTGEVTP	93.9943	−6.4550	−6.5826	8.2616

correlation between general trends of every pair of variables. If they show an increasing trend simultaneously, the result will be a positive correlation coefficient.

A Fisher-type test was applied to check the stationarity of all variables. In the context of panel data, the Fisher-type test performs a unit root test for each panel individually and then combines the p-values from these tests to obtain an overall test to identify whether the panel series contains a unit root. The null hypothesis tested by the Fisher-type test is that all panels contain a unit root. For a finite number of panels, the alternative hypothesis is that at least one panel has stationarity.

Table 7.3 shows the results for Fisher-type unit root tests for all panels based on augmented Dickey–Fuller tests. All tests rejected the null hypothesis, and it is detected that all panels are stationary.

7.3.2   Model Specification and Testing Functional Form

As previously mentioned, the standard form of the EKC is defined as Eq. (7.4). In order to evaluate the impact of renewable energy deployment, technological innovation in energy applications adopted with climate change mitigation, and ICTs,

we include the additional variables related to these parameters and test the functional form by comparing it with two quadratic equations in the form of the log-linear functions as follows:
Model 1:

$$\ln{(CO_2/P)_{it}} = \alpha_i + \beta_1 \ln{(GDP/P)_{it}} + \beta_2 \left(\ln{(GDP/P)^2}\right)_{it} + \varepsilon_{it} \qquad (7.4)$$

Model 2:

$$\ln{Y_{it}} = \alpha_0 + \sum_j \beta_j \ln{X_{jit}} + \frac{1}{2} \sum_j \beta_{jj} \ln{X_{jit}^2} + \alpha_t t + \frac{1}{2} \alpha_{tt} t^2 + \varepsilon_{it} \qquad (7.7)$$

Model 3:

$$\ln{Y_{it}} = \alpha_0 + \sum_j \beta_j \ln{X_{jit}} + \frac{1}{2} \sum_j \beta_{jj} \ln{X_{jit}^2} + \sum_i \sum_j \beta_{jk} \ln{X_{jit}} \ln{X_{kit}}$$
$$+ \alpha_t t + \alpha_{tt} t^2 + \sum_j \beta_{jt} \ln{X_{jit}} t + \varepsilon_{it}, \qquad (7.6)$$

These models are frequently used by researchers. The translog functional form (Model 3) has become more popular because it provides more flexibility in computation of effects (Corbo and Meller 1979). It accounts for nonlinearity by using both squares and interactions of the explanatory variables.

We employed the likelihood ratio (LR) test to select the best functional form. The likelihood ratio is defined as follows (Johnston and DiNardo 2007):

$$\lambda = \frac{L\left(\tilde{\beta}, \tilde{\sigma}^2\right)}{L\left(\widehat{\beta}, \widehat{\sigma}^2\right)},$$

where the result value of $L\left(\widehat{\beta}, \widehat{\sigma}^2\right)$ is the unrestricted maximum likelihood and $L\left(\tilde{\beta}, \tilde{\sigma}^2\right)$ is related to restricted function. We estimated these models and derived the log-likelihood value in order to calculate the LR test for each functional form. It was expected that the null hypothesis would be rejected if λ were small. The LR test could be calculated as follows:

$$\text{LR} = -2 \ln{\lambda} = 2 \left[\ln{L\left(\widehat{\beta}, \widehat{\sigma}^2\right)} - \ln{L\left(\tilde{\beta}, \tilde{\sigma}^2\right)}\right] \sim \chi^2_{(q)}. \qquad (7.8)$$

The results of the LR test for all three functions are presented in Table 7.4, which shows that the translog functional form (Model 3) is an appropriate equation to estimate the relation. Here, q is the number of parameter restrictions.

We employed the Akaike information criterion (AIC) and the Bayesian information criterion (BIC) to compare the functions for model specification. The AIC and

Table 7.4 LR test for functional form

Function	LR test	Critical value	Result
Model 3 vs Model 1	340.00976	14.611	Model 3 accepted
Model 2 vs Model 1	221.13196	3.940	Model 2 accepted
Model 3 vs Model 2	118.87780	1.145	Model 3 accepted

Table 7.5 Model specification test

Model	Obs.	Ln(L)	df	AIC	BIC
Model 1	240	45.3051	18	−54.6102	8.0412
Model 2	240	155.8711	28	−255.7423	−158.2844
Model 3 (TL)	240	215.3100	43	−344.6200[**]	−194.9525[**]

Note: [**] indicates significance at 5% level.

BIC are two popular statistical measures to compare models. They are defined as follows:

$$AIC = -2 * \ln(\text{likelihood}) + 2 * k \qquad (7.9)$$

$$BIC = -2 * \ln(\text{likelihood}) + \ln(N) * k, \qquad (7.10)$$

where k is the number of parameters estimated and N is the number of observations. AIC and BIC are measures that combine fit and complexity. Fit is measured negatively by $-2*\ln(\text{likelihood})$; the larger the value, the worse the fit. Complexity is measured positively, either by $2*k$ for AIC or $\ln(N)*k$ for BIC. Because the two models fit on the same data, the model with the smaller value of information criterion is considered better (*Stata Base Reference Manual*, 2012). The results of AIC and BIC for comparing models are presented in Table 7.5.

Based on the results, the model specified by the translog function (Model 3) is selected as the appropriate model specification because it has the smallest value in both AIC and BIC measurements.

7.3.3   Estimation Method, Testing, and Selection of Final Method

This model is usually estimated using panel data. Most studies have estimated both fixed-effect and random-effect models. As previously mentioned Eq. (7.1), the fixed-effect model assumes that α_i and γ_t are correlated with explanatory variables, but they are considered error components in the random-effect model. The least-squares dummy variable (LSDV) is then applied to estimate the fixed model. In the case of the random-effect model, the residuals form and the OLS are used to construct a variance–covariance matrix in order to estimate the model by the feasible generalized least squares (FGLS) technique. If a correlation is found between α_i, γ_t, and the explanatory variables, the random-effect model is inconsistent, and only the fixed-effect model can estimate the regression consistently (Mundlak 1978; Hsiao 2003). A Hausman (1978) test was used to compare the fixed-effect and random-effect models in terms of their consistency. With the assumption that no other statistical problem exists, the fixed-effect model is estimated consistently.

Our estimation method differs from most studies because it uses feasible generalized least squares (FGLS) to correct for heteroscedasticity and autocorrelation. Although heteroscedasticity and autocorrelation in residuals are sources of serious problems, none of the early studies presented diagnostic tests of models. Stern et al. (1996) identified that the heteroscedasticity problem is important when we are concerned with cross-sectionals of grouped data. Many researchers (Cropper and Griffiths 1994; Shafik 1994; Horvath 1997; Moomaw and Unruh 1997; Suri and Chapman 1998) estimated fixed-effect models without presenting regression diagnostic tests. Stern (2002) estimated a decomposition of EKC using an FGLS model. Suri and Chapman (1998), Aldy (2005), and Luzzati and Orsini (2009) obtained results employing the FGLS approach in order to correct cross-sectional heteroscedasticity and serial correlation.

Based on the literature, autocorrelation and heteroscedasticity in panel data models will cause the results achieved by the models to be biased and less efficient. We, therefore, need to check these problems in our model. Stata (*Stata Quick Reference and Index* 2012) implements a test for serial correlation in the idiosyncratic errors of a linear panel data model discussed by Wooldridge (2002). Drukker (2003) presented simulation evidence that this test has good size and power properties in a reasonable sample size. When we applied the Wooldridge test for serial correlation, the null hypothesis of no first-order autocorrelation was rejected. The result is presented in Table 7.6. As the table shows, the computed value for the F test (33.107) exceeds the critical F value obtained from the F table at the chosen level of significance (or the probability value).

This test provides an important alternative to the Wald testing for models by maximum likelihood. Because the Wald testing requires fitting only one model (the unrestricted model), it is computationally more attractive than the likelihood ratio testing. Hence, it is used whenever feasible because the null distribution of the LR test statistics is often more closely chi-squared distributed than the Wald test statistics are (Stata Quick Reference and Index 2012).

Since iterated GLS with only heteroscedasticity generates the maximum likelihood parameter, we are able to calculate an LR test by comparing the estimation of a model fitted with panel-level heteroscedasticity and a model without heteroscedas-

Table 7.6 Serial correlation test result

Wooldridge test for autocorrelation in panel data
H0: no first-order autocorrelation
$F(1,14) = 33.1070$
Prob $> F = 0.0000$

Table 7.7 Heteroscedasticity test result

Likelihood-ratio test
Assumption: homoscedasticity nested in heteroscedasticity
LR chi-square (14) = 152.1300
Prob > F = 0.0000

ticity. Based on the result, the null hypothesis is rejected. Therefore, the presence of heteroscedasticity is detected, as shown in Table 7.7. The computed value for the LR test (152.13) exceeds the critical chi-square value obtained from the F table at the chosen level of significance (or the probability value).

Thus, in the presence of heteroscedasticity and serial autocorrelation, we applied FGLS to estimate our model. It is asymptotically more efficient than OLS and other estimators (Wooldridge 2002).

7.3.4 Estimation Result

The results of our estimation reported in Table 7.8 are as follows. The share of coefficients, which were estimated with a high significance level ($p < 0.001$), is greater than 60 %. If we consider a general significance level of 0.05, it will be more than 70 %. The variables are as follows: CO_2 per capita (cdecap), GDP per capita (gdpcap), contributions of renewable energy sources in total power generation (regenp), number of energy patents per million inhabitants (ptgcap), number of ICT patents per million inhabitants (pticap), environmental tax per capita (evtcap), and trend. All variables have been considered logarithmic values. The signs related to renewable energy generation, energy patents, and environmental tax are negative, as we expected. The sign for the linear and square of energy patents is negative, but the second one is not statistically significant.

Considering the negative sign for the first order of GDP and the positive sign for the second order, the relation between CO_2 emissions and GDP shows a convex curve for the EU-15 countries (Fig. 7.1). Regarding GDP, our result contrasts with early studies focused on the EKC that argued for the inverted U-shaped curve relationship between emissions per capita and GDP per capita in developed countries. The U-shaped curve might be due to the fact that our sample consists of only advanced economies where all countries have reached the turning points and an additional increase in their welfare can be achieved only at higher environmental cost. According to the literature, the relationship between environmental quality, level of income, and other variables was tested and analyzed by several researchers (Shafik and Bandyopadhyay 1992; Panayotou 1993, 1997; Selden and Song 1994; Torras and Boyce 1998; Suri and Chapman 1998). Based on their findings, there may be willingness to accept a weak level of environmental quality in the early stage of development, but a turning point will be achieved as the level of income increases.

Table 7.8 Feasible Generalized Least Squares (FGLS) estimation result

Variables	Model 1 Coeff.	P > |z|	Model 2 Coeff.	P > |z|	Model 3 Coeff.	P > |z|
LnGDP/P	−1.14808	0.000	−2.01421	0.000	−0.32725	0.328
(LnGDP/P)^2	0.20777	0.000	0.32874	0.000	0.28970	0.000
LnRen/TPG			−0.02128	0.102	−0.26618	0.001
(LnRen/TPG)^2			0.02103	0.000	0.04580	0.000
LnPateng/P			−0.01032	0.000	−0.40697	0.000
(LnPateng/P)^2			−0.00307	0.003	−0.00321	0.191
LnPatict/P			0.02531	0.000	1.03926	0.000
(LnPatict/P)^2			−0.00542	0.000	−0.00114	0.409
LnEvt/P			0.12302	0.000	−1.83548	0.000
(LnEvt/P)^2			−0.04155	0.000	−0.17791	0.000
Trend			0.03569	0.000	0.18498	0.000
Trend^2			−0.00248	0.000	−0.00192	0.000
(LnGDP/P)(LnRen/TPG)					0.09861	0.000
(LnGDP/P)(LnPateng/P)					0.10013	0.000
(LnGDP/P)(LnPatict/P)					−0.28603	0.000
(LnGDP/P)(LnEvt/P)					0.37623	0.000
(LnGDP/P)Trend					−0.04738	0.000
(LnRen/TPG)(LnPateng/P)					−0.00773	0.039
(LnRen/TPG)(LnPatict/P)					0.00437	0.262
(LnRen/TPG)(LnEvt/P)					−0.01784	0.389
(LnRen/TPG)Trend					−0.00917	0.000
(LnPateng/P)(LnPatict/P)					0.00372	0.255
(LnPateng/P)(LnEvt/P)					−0.09005	0.000
(LnPateng/P)Trend					0.00089	0.231
(LnPatict/P)(LnEvt/P)					0.19941	0.000
(LnPatict/P)Trend					−0.00236	0.002
(LnEvt/P)Trend					0.02046	0.000
Const.	3.67216	0.000	5.05640	0.000	−0.30706	0.625
Number of obs	240		240		240	
Number of groups	15		15		15	
Time periods	16		16		16	
Wald chi2	516.71		1,389.29		2,549.14	
Prob > chi2	0.0000		0.0000		0.0000	

However, this concept was challenged recently by researchers such as Harbaugh et al. (2002) and Millimet et al. (2003). Wagner (2008) claimed that the evidence of an inverted U-shaped relationship between CO_2 emissions and GDP obtained with commonly used methods is entirely spurious because of several major econometric problems. In line with this approach, we cast doubt on the results achieved by previous researchers because they cannot be confirmed on the basis of our findings. As Dasgupta et al. (2004) pointed out, the less robust relationship between GDP

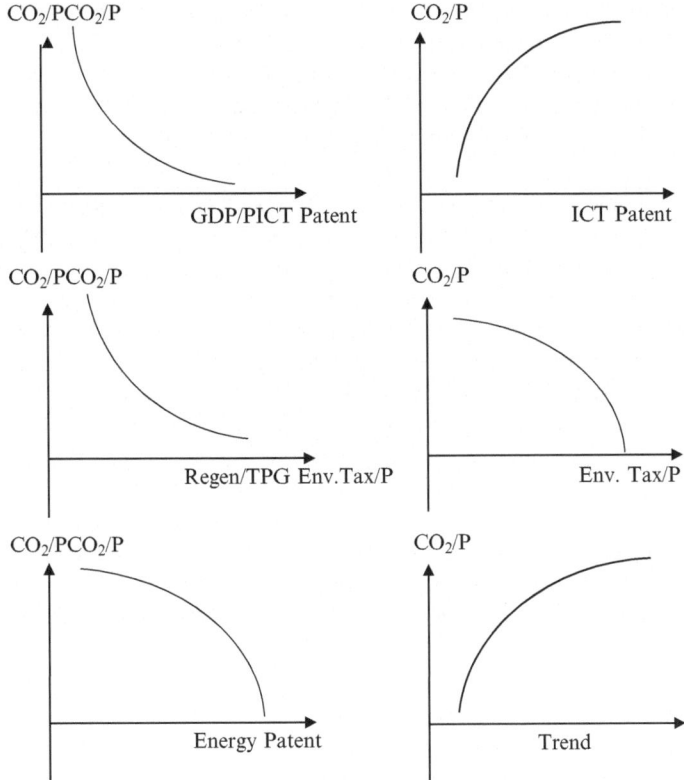

Fig. 7.1 The relationship between explanatory variables and CO_2 emissions

and CO_2 emissions caused some econometric critiques; therefore, this relationship is not rigid as found by previous researchers. Moreover, the role of GDP growth in CO_2 emission reduction could be affected by the governance-related explanatory variables.

Figure 7.1 shows an increase in the decreasing rate of CO_2 emissions in relation to the variables of energy patents and environmental tax. Furthermore, considering that we estimated our model from 1995 to 2010, and the financial crisis occurred within this period, the EU countries had to reduce expenditures for renewable energy and coal-fueled power plants restarted operations because of the lower unit cost compared to imported crude oil and natural gas. In addition, the recession in the early 2000s should be taken into account. It was a downtrend in economic activity, which mainly happened in developed countries. The EU was affected by this recession from 2000 to 2002. Therefore, all indicators, including GDP, number of patents, taxes, and CO_2 emissions, could be affected by the recession in the early 2000s and the financial crisis from 2008 to 2009.

The relationship between variables and CO_2 emissions per capita is presented in Fig. 7.1. The figure shows a concave formation between CO_2 emissions per capita (CDECAP) with GDP per capita (GDPCAP) and the share of renewable energy sources in total power generation (REGENP). It defines a decrease in the decreasing rate of CO_2 emissions regarding GDP and renewable energy in the EU-15. The results also showed a convex formation between CO_2 emissions per capita (CDECAP), the number of ICT patents per million inhabitants (PTICAP), and technological change over time (TREND). Therefore, the results showed a decrease in the increasing rate of CO_2 emissions in relation to ICT patents and technological change over time. Furthermore, there is a convex formation between CO_2 emission per capita (CDECAP), energy patent applications per million inhabitants (PTGCAP), and environmental tax per capita (EVTCAP). The results showed an increase in the decreasing rate of CO_2 emissions in relation to the variables of energy patents and environmental tax.

ICT patents showed relevant positive and negative effects on CO_2 emissions in different ways: the positive effects were increased electronic wastes, and the negative effects were improved energy efficiency. The impact of ICT on emissions was closely related to energy consumption (Hilty et al. 2006). The main increase in effects was caused by freight transport and ICT's demand for electricity in manufacturing and the disposal of hardware. Energy saving is made by virtual goods, whereas ICT supported management and control of the production process. Regarding the relationship between ICT patents and CO_2 emissions, the results showed higher ICT growth with lower emission increases over time. Our findings are consistent with Romm (2002), who argued that recent reductions in energy intensity were related to IT growth, which, being less energy intensive, increased efficiency in other sectors of the economy.

The results showed that CO_2 emissions are reduced through power generated by renewable energy sources. The impact of renewable energy deployment on CO_2 emission reduction has been discussed in Chap. 3. As previously mentioned, there is extensive literature regarding the potential of carbon saving using renewable energy technologies. However, the reductions made by renewable energy enhancement have decreased from 1995 to 2010 in the EU-15. The employment trend in renewable energy has been affected worldwide by global recession, policy changes, and overcapacities in the wind and solar supply chains (Martinot and Sawin 2012). Therefore, the decreasing rates in the effectiveness of renewable energy could be explained by the reduction in the growth rate of renewable energy deployment.

Based on our findings, the impact of environmental tax on CO_2 emissions has increased over time. This is consistent with the results obtained by Alfsen et al. (1995) regarding the sizable effects of carbon tax on emissions in Western Europe. They indicated that external benefits, such as reduction in health damage, damage to nature, and road traffic, are associated with carbon and energy taxes. Our findings are also consistent with Brännlund and Nordström (2004), who studied the effects of CO_2 tax in Sweden. They found that the demand for all fossil fuel–related goods was decreasing because of the CO_2 tax.

As shown in Fig. 7.1, the number of patents of energy technology for mitigation or adaptation to climate change had a similar impact on CO_2 emissions compared to environmental tax. This is consistent with Popp (2001, 2005), who argued that it is important to consider the role of technological innovation in considering solutions to long-term environmental problems, such as energy consumption and climate change. Furthermore, according to Nordhaus (2002), it is expected that induced innovation would lead to a reduction in carbon intensity and the cumulative effect might be larger in the long term.

Time trend showed a decrease in the increasing rate of CO_2 emissions per capita. It indicated that the growth rate of CO_2 emissions is reduced over time. The amount of CO_2 is increasing because the amount of production and consumption is increasing. However, the rate is decreasing because of the effects of technological change, productivity, and efficiency. This finding is consistent with Popp (2005) regarding the gradual process of the diffusion and adoption of new technologies. It is also consistent with the implications of energy efficiency technologies for climate policy, as discussed by Jaffe et al. (2001, 2003).

An alternative model was estimated using environmental tax divided by fossil fuel consumption as a proxy for tax rate. The results showed that the coefficients for ICT patents were positive, which is not acceptable because it indicates that the impact of ICT technologies on CO_2 emissions is positive and it is increasing. Furthermore, the results of the environmental tax rate based on fossil fuel consumption showed that its marginal effectiveness is decreasing with increase in the ratio of environmental tax to fossil fuel consumption, which is not reasonable because this is considered an inelastic parameter. Energy consumption will be more sensitive to environmental tax when it passes a certain level. The result is shown in the Appendix Tables A.1, A.2, and A.3.

A normalized model with the mean of variables was also tested. The results showing impact of renewable energy generation and environmental tax were not consistent with expectations. Based on the estimation, there is an increasing rate of CO_2 emissions when the power generated by renewable energy sources increases. CO_2 emissions increase at a decreasing rate when environmental tax increases. This result is presented in the Appendix Tables B.1, B.2, and B.3.

7.3.5 Elasticities

Although the parameters in the translog function do not have a direct interpretation, they show the formation of a complex and nonlinear relationship between dependent variables and explanatory variables. Considering that our model was estimated in logarithmic form, we can calculate the overall elasticity of CO_2 emissions to GDP, renewable energy generation, and environmental tax. These elasticities measure the percentage change in emissions caused by a percentage change in each variable, everything else given. Based on Eq. (7.6), the formula used to calculate elasticities and rate of technical change is as follows:

$$E_{jit} = \delta \ln Y_{it}/\delta \ln X_{jit} = \beta_j + \beta_{jj} \ln X_{jit} + \sum \beta_{jk} \ln X_{kit} + \beta_{jt}t \quad (7.11)$$

$$\text{TC}_{it} = \delta \ln Y_{it}/\delta t = \beta_t + \beta_{tt}t + \sum \beta_{jt} \ln X_{jit} \qquad (7.12)$$

Based on the results, the elasticity of carbon dioxide emission for the EU-15 countries over the period of 1995–2010 is calculated as 0.238, −0.165, and 0.267 for GDP, renewable energy generation, and environmental tax respectively.

The results showed that the elasticity for each variable is affected by interaction with the other variables because of the interactive relationship between the parameters. Therefore, elasticity is positive for environmental tax, whereas its coefficient is negative and strongly significant in our model. Generally, we face this situation when the translog function is applied. If model 1 is employed to estimate the model, we could have expected a given sign for the variables. However, this did not occur in the translog function because of the interaction relationship between the explanatory variables and the square terms' coefficients. The advantage of the translog function, compared with the other models, is its ability to evaluate these interaction effects.

In addition to interaction effects, the positive sign of environmental tax could be explained by the level of tax and its impact on consumer revenue. The EU-15 countries are considered developed countries, and their GDP per capita is relatively high. Therefore, the tax rate should be high enough to cause a sensitive reduction in energy consumption. This finding is consistent with Pearce (1991), who showed that a reduction in CO_2 emissions might not be achieved unless the related elasticities are known with acceptable certainty. Moreover, Howarth (2006) pointed out that private consumption generates negative externality by increasing the standard of living. Hence, we can argue that environmental tax has a negative effect on CO_2 emissions indirectly by making an incentive to enhance renewable energy development or create technological innovation. Tax revenues could cause an increase in household energy consumption through the revenue recycling effect, in which revenues are used to reduce the tax rate on income or provide increased transfer payments to households. .

7.3.6   Technological Change

Regarding the elasticity of CO_2 emissions in relation to energy patent applications, the results showed that ICT patents and trends were −0.0249, 0.005, and −0.002 respectively. Our findings are consistent with the literature regarding the links between environmental regulation, innovation, and technological change (Lanjouw and Mody 1996; Buonanno et al. 2003; Popp 2006). The elasticities showed that energy patent applications that adopted climate change mitigation caused a reduction of 0.02 % in CO_2 emissions per capita for an increase of 1.0 % in the number of patents per million inhabitants. Elasticity for trend, considered as the rate of technological change, defined as percentage reduction in CO_2 over time, everything else given, implies a negative, small impact on CO_2 emissions. The elasticity of ICT patents was positive and small, around 0.01, which can

Table 7.9 CO_2 elasticities in the EU-15 countries over 1995–2010

	GDP	Regenp	Ptgcap	Pticap	Evtcap	Trend
1995	0.6171	−0.1133	−0.0173	0.0010	0.0740	0.0284
1996	0.5471	−0.1220	−0.0204	0.0077	0.1045	0.0251
1997	0.3783	−0.1300	−0.0226	0.0160	0.1530	0.0225
1998	0.3058	−0.1301	−0.0241	0.0190	0.1997	0.0178
1999	0.1929	−0.1443	−0.0244	0.0214	0.2539	0.0146
2000	0.0436	−0.1569	−0.0234	0.0219	0.2779	0.0122
2001	−0.0370	−0.1697	−0.0243	0.0241	0.3014	0.0092
2002	−0.0195	−0.1920	−0.0237	0.0214	0.3074	0.0062
2003	0.1044	−0.1845	−0.0235	0.0083	0.3322	−0.0025
2004	0.2021	−0.1784	−0.0258	0.0025	0.3301	−0.0095
2005	0.1798	−0.1850	−0.0269	0.0010	0.3396	−0.0139
2006	0.2031	−0.1854	−0.0281	−0.0043	0.3320	−0.0195
2007	0.3039	−0.1830	−0.0272	−0.0165	0.3267	−0.0265
2008	0.3250	−0.1794	−0.0274	−0.0224	0.3337	−0.0325
2009	0.2090	−0.1906	−0.0303	−0.0107	0.3233	−0.0335
2010	0.2558	−0.1920	−0.0286	−0.0063	0.2792	−0.0356
Mean	0.2382	−0.1648	−0.0249	0.0053	0.2668	−0.0023
Std dev	0.1793	0.0277	0.0033	0.0147	0.0869	0.0223

be considered an effective parameter in climate change mitigation. This result is consistent with Fuchs (2008) and confirms that the ICT sector emits less CO_2 than the total economy. Fuchs indicated that the ICT sector constitutes a small portion of the total value added and the most dominant economic activity in modern industrialized economies is fossil fuel combustion. The elasticities, as we will see in the next section, confirm this viewpoint, and the numbers vary across the countries in the EU-15.

7.3.7 Variations in Elasticities Over Time

The variation in elasticities of CO_2 emissions per capita in the EU-15 countries over time are presented in Table 7.9. The time trend shows that the effectiveness of renewable energy generation, innovation, and technological change on CO_2 emission reduction has improved steadily. There is a negative elasticity for GDP from 2001 to 2002, which could be explained by the recessions in Europe in the early 2000s.

The results showed that the elasticity of CO_2 emission per capita for tax has an upward trend, which is consistent with previous comments about the effects of environmental tax, interaction, and revenue recycling. The general trend of elasticities for renewable energy, energy patents, ICT patents, and trend steadily shows increased effectiveness in climate change mitigation. The elasticities of two

Table 7.10 CO_2 elasticities in the EU-15 by country

	GDP	Regenp	Ptgcap	Pticap	Evtcap	Trend
Austria	0.3186	−0.0187	−0.0178	−0.0373	0.3358	−0.0228
Belgium	−0.1164	−0.3010	0.0187	−0.0697	0.4787	0.0042
Germany	0.8316	−0.2425	−0.2009	0.3898	−0.3302	0.0389
Denmark	0.7046	−0.1485	−0.0666	0.0620	0.0807	−0.0032
Spain	0.1419	−0.1556	0.0161	−0.0659	0.1926	−0.0033
Finland	−0.0791	−0.0948	−0.0275	0.0048	0.5251	−0.0127
France	−0.0662	−0.1719	0.0119	−0.0721	0.4778	−0.0092
Greece	0.2423	−0.2387	−0.0001	−0.0090	−0.0552	0.0150
Ireland	0.2294	−0.2107	0.0123	−0.0728	0.4247	−0.0066
Italy	0.2449	−0.1474	−0.0217	0.0054	0.1975	−0.0014
Luxembourg	1.0530	−0.0381	−0.0156	−0.0822	0.2663	−0.0322
Netherlands	0.0437	−0.2401	−0.0327	0.0270	0.4054	0.0041
Portugal	0.4437	−0.1417	−0.0286	0.0463	−0.2309	0.0122
Sweden	0.1644	−0.0433	−0.0281	−0.0128	0.4528	−0.0205
UK	−0.5830	−0.2787	0.0075	−0.0346	0.7806	0.0023
Mean	0.2382	−0.1648	−0.0249	0.0053	0.2668	−0.0023
Std dev	0.4078	0.0883	0.0541	0.1158	0.2996	0.0172

alternative models were calculated, and the results are presented in the Appendix. In both models, the sign of environmental tax elasticity is negative. However, the sign of renewable energy is positive in the normalized model.

7.3.8 Variations in Elasticities by Country

Table 7.10 presents the elasticities of CO_2 emissions per capita to GDP per capita, share of electricity generated by renewable energy, energy patent applications per million inhabitants, ICT patents per million inhabitants, environmental tax per capita, and trend in the EU-15 from 1995 to 2010 by country. The elasticities vary across countries, which can be explained by different demography, geography, economic structure, and policies of the member countries.

7.4 Examples of Policy Implications for Developing Countries

Our findings could be used by countries that attempt to develop ecocities, such as Dongtan City in China and Masdar City in the United Arab Emirates (UAE). Masdar City is supposed to rely entirely on solar energy and other renewable energy sources. Because of the wide range of strategies in different countries, which we have already analyzed, there is a good opportunity for the UAE government to take advantage of our findings to formulate policies to promote renewable energy in Masdar City. It is possible for the government to apply efficient energy policies at the starting point. The buildings of Masdar City are going to utilize energy-efficient construction material (Premalatha et al. 2013). Targets must be set, and policies should be formulated to achieve these targets. For example, although European countries applied FIT policies to develop renewable energy sources, they are now trying to build a harmonized market for renewable energy. However, FIT is not an appropriate policy for this purpose.

Al-Amir and Abu-Hijleh (2013) investigated the strategies used in different countries in order to determine the best practical policy to apply in the UAE. As we pointed out, it is necessary to consider the effects of different variables on CO_2 emission reduction and the effects of their interactions. In other words, we should use a top–down approach to design a policy for climate change mitigation. We can then apply a bottom–up approach to select competitive technology in order to achieve the target. According to Al-Amir and Abu-Hijleh, the government should apply a combination of FIT, RPS, and tax incentive policies. Based on our findings, all elasticities should be known and reliable in order to formulate an effective policy. As already analyzed, a tax policy will be not effective if it does not set taxes high enough to provide incentives to eliminate fossil fuel products.

Masdar City has an advantage that differentiates it from other ecocities. It has the potential to establish the first market for trading renewable energy, which we discussed in Chap. 6. Partial trading has been already applied in Dubai, and traders are familiar with this mechanism. Therefore, a similar mechanism could be applied to trading commoditized renewable energy in small amounts. In addition, the microcertificate could be applied as a type of partial trading for the RPS mechanism in this market. Considering that ICT infrastructure and advanced technologies are supposed to be used in the building of Masdar City, the basis for creating a new energy market is available.

7.5 Summary of the Effects of Renewable Energy Development on CO_2 Reduction

In this chapter, we evaluated the impact of renewable energy development on CO_2 emission reduction. We also investigated the effectiveness of innovation in energy technologies adopted to reduce CO_2 emissions, ICT technologies, and environmental taxes applied to encourage renewable energy development. Environmental tax was considered in the model in order to evaluate the effects of market regulation. Several scholars have applied different methodologies to examine the relationship between energy consumption and economic growth in individual countries and in groups of countries in order to analyze the effects of governmental energy policies. Many researchers have studied the relationship between CO_2 emissions and economic growth by using the EKC. These studies analyzed variables such as population, inequality, trade, and openness. Most results showed that environmental quality would be promoted after a certain level of economic growth was achieved. Therefore, developed countries with a high level of GDP per capita would promote environmental quality.

Recently, this idea has been challenged by theoretical and econometric critiques. Recent studies have shown that the methodology used by previous researchers was not appropriate to estimate this relationship. Moreover, the important role of governance-related variables was neglected. Our contribution to the literature is to add electricity generated by renewable energy sources, technological innovation in energy technology patents, ICT, and environmental tax to the model. All these variables are considered governance related, either directly or indirectly. Environmental tax applied by governments is an example of the direct effect of governance. On the other hand, CO_2 emissions are indirectly affected by governmental policy through technological innovation and renewable energy generation. Furthermore, we applied the FGLS method to estimate the model in order to avoid major econometric problems.

In contrast with previous research, we found that economic growth might not lead to promotion of environmental quality. The results showed a positive relation between CO_2 emissions per capita and GDP per capita in the EU-15, which is comprised of developed countries. Previous research found a negative relation between these variables. In popular myths about the effects of ICT technological innovation and environmental tax, both factors negatively affect CO_2 emissions. However, our results showed that the impact of ICT differs across countries based on their structure. Environmental tax had a negative and strong effect on CO_2 emissions by itself, but the positive elasticity showed that its negative effect becomes positive because of the high-income and revenue-recycling effects of tax policy.

In summary, we consider that the role of governmental policymaking is more important than economic growth. Renewable energy sources have been promoted in the EU-15 because of governmental support mechanisms and subsidies. Environmental tax policy and tightened standards could lead to more patents. Therefore, CO_2 emissions will be more affected by governmental parameters than by achieving a certain level of economic development. This result could be important for the climate change policies of developing countries. In other words, it is not necessary to obtain a high level of economic growth in order to enhance a country's environmental quality. Developing countries are able to achieve this target through appropriate policymaking by their governments. Hence, developing countries could achieve high levels of environmental quality before achieving high levels of GDP per capita.

References

Akarca AT, Long TV (1980) Relationship between energy and GNP: a reexamination. J Energy Develop 5(2):326–331

Al-Amir J, Abu-Hijleh B (2013) Strategies and policies from promoting the use of renewable energy resource in the UAE. Renew Sustain Energy Rev 26:660–667. http://dx.doi.org/10.1016/j.rser.2013.06.001

Aldy JE (2005) An environmental Kuznets curve analysis of US state-level carbon dioxide emissions. J Environ Develop 14(1):48–72

Alfsen KH, Birkelund H, Aaserud M (1995) Impacts of an EC carbon/energy tax and deregulating thermal power supply on CO2, SO2 and NOx emissions. Environ Resour Econ 5(2):165–189

Allen EL, Cooper CL, Edmonds FC, Edmonds JA, Reister DB, Weinberg AM, Whittle CE, Zelby LW (1976) US energy and economic growth, 1975–2010, Institute for Energy Analysis, Oak Ridge

Amemiya T (1985) Advanced econometrics. Harvard University Press, Cambridge, MA

Baltagi B (2008) Econometric analysis of panel data, 4th edn. Wiley-Blackwell, Chichester

Benitez LE, Benitez PC, Van Kooten GC (2008) The economics of wind power with energy storage. Energy Economics 30(4):1973–1989

Bhattacharyya SC (2011) Energy economics: concepts, issues, markets and governance. Springer, London

Böhringer C (1998) The synthesis of bottom-up and top-down in energy policy modeling. Energy Economics 20(3):233–248

BP (2012) BP Statistical Review of World Energy. www.bp.com/statisticalreview/

Brännlund R, Nordström J (2004) Carbon tax simulations using a household demand model. Eur Econ Rev 48(1):211–233

Buonanno P, Carraro C, Galeotti M (2003) Endogenous induced technical change and the costs of Kyoto. Resour Energy Econ 25(1):11–34. http://dx.doi.org/10.1016/S0928-7655(02)00015-5

Cameron AC, Trivedi PK (2009) Microeconometrics using STATA, vol 5. Stata Press, College Station

Cheng BS (1995) An investigation of cointegration and causality between energy consumption and economic growth. J Energy Dev 21(1):73–84

Choi E, Heshmati A, Cho Y (2010) An empirical study of the relationships between CO2 emissions, economic growth and openness. IZA, Bonn

Cochrane D, Orcutt GH (1949) Application of least squares regressions to relationships containing autocorrelated error terms. J Am Stat Assoc 44:32–61

Cole MA, Rayner A, Bates JM (1997) The environmental Kuznets curve: an empirical analysis. Environ Dev Econ 2(04):401–416

Coondoo D, Dinda S (2002) Causality between income and emission: a country group-specific econometric analysis. Ecol Econ 40(3):351–367. http://dx.doi.org/10.1016/S0921-8009(01)00280-4

Corbo V, Meller P (1979) The translog production function: some evidence from establishment data. J Econ 10(2):193–199. http://dx.doi.org/10.1016/0304-4076(79)90004-6

Cropper M, Griffiths C (1994) The interaction of population growth and environmental quality. Am Econ Rev 84(2):250–254

Dasgupta S, Hamilton K, Pandey K, Wheeler D (2004) Air pollution during growth: accounting for governance and vulnerability. World Bank Policy Research Working Paper 3383

Depuru SSSR, Wang L, Devabhaktuni V (2011) Smart meters for power grid: challenges, issues, advantages and status. Renew Sust Energ Rev 15(6):2736–2742

Drukker DM (2003) Testing for serial correlation in linear panel-data models. Stata J 3(2):168–177

EIA U.S. What drives crude oil prices? U.S. Energy Information Administration. Retrieved 18 Apr 2013, from http://www.eia.gov/finance/markets/

Faruqui A, Hledik R, Newell S, Pfeifenberger H (2007) The power of 5 percent. Electr J 20(8):68–77

Ford A (1995) The impacts of large scale use of electric vehicles in Southern California. Energy Build 22(3):207–218

Frankl P, Masini A, Gamberale M, Toccaceli D (1997) Simplified life-cycle analysis of PV systems in buildings: present situation and future trends. INSEAD, Centre for the Management of Environmental Resources

Fuchs C (2008) The implications of new information and communication technologies for sustainability. Environ Dev Sustain 10(3):291–309. doi:10.1007/s10668-006-9065-0

Galinato GI, Yoder JK (2010) An integrated tax-subsidy policy for carbon emission reduction. Resour Energy Econ 32(3):310–326

Grossman GM, Krueger AB (1991) Environmental impacts of a North American free trade agreement, National Bureau of Economic Research, NBER Working Paper No. 3914

Grubb M, Edmonds J, Ten Brink P, Morrison M (1993) The costs of limiting fossil-fuel CO2 emissions: a survey and analysis. Annu Rev Energy Environ 18(1):397–478

Harbaugh WT, Levinson A, Wilson DM (2002) Reexamining the empirical evidence for an environmental Kuznets curve. Rev Econ Stat 84(3):541–551

Hartway R, Price S, Woo C (1999) Smart meter, customer choice and profitable time-of-use rate option. Energy 24(10):895–903

Hausman JA (1978) Specification tests in econometrics. Econometrica J Econ Soc 46:1251–1271

Hilty LM, Arnfalk P, Erdmann L, Goodman J, Lehmann M, Wäger PA (2006) The relevance of information and communication technologies for environmental sustainability – a prospective simulation study. Environ Model Softw 21(11):1618–1629. http://dx.doi.org/10.1016/j.envsoft.2006.05.007

Hitch CJ (1978) Energy conservation and economic growth, AAAS Selected Symposium 22. Published by Westview Press for the American Association for the Advancement of Science, Boulder

Horvath RJ (1997) Energy consumption and the environmental Kuznets curve debate. Department of Geography, University of Sydney, Sydney

Howarth RB (2006) Optimal environmental taxes under relative consumption effects. Ecol Econ 58(1):209–219

Hsiao C (2003) Analysis of panel data, vol 34, 2nd edn. Cambridge University Press, Cambridge

Huang BN, Hwang M, Yang C (2008) Causal relationship between energy consumption and GDP growth revisited: a dynamic panel data approach. Ecol Econ 67(1):41–54

IEA (2012) CO2 emissions from fuel combustion 2012. OECD Publishing, Paris

Jackman S (2004) Generalized least squares, Stanford University. http://jackman.stanford.edu/papers/gls.pdf. Retrieved on 1 Nov 2014

Jaffe AB, Palmer K (1997) Environmental regulation and innovation: a panel data study. Rev Econ Stat 79(4):610–619

Jaffe AB, Newell RG, Stavins RN (2001) Energy-efficient technologies and climate change policies. In: Toman M (ed) Climate change economics and policy. An RFF anthology. RFF Press, Washington, DC, pp 171–181

Jaffe AB, Newell RG, Stavins RN (2003) Technological change and the environment. Handb Environ Econ 1:461–516

Jansen B, Binding C, Sundstrom O, Gantenbein D (2010) Architecture and communication of an electric vehicle virtual power plant. Paper presented at the Smart Grid Communications (SmartGridComm), 2010 First IEEE International Conference on, IBM Research-Zurich, Switzerland

Johnston J, DiNardo J (2007) Econometric methods, vol 4. Wiley Online Library, McGraw-Hill Education, Singapore

Judge GG, Hill RC, Griffiths W, Lutkepohl H, Lee TC (1988) Introduction to the theory and practice of econometrics. Willey, New York

Kempton W, Letendre SE (1997) Electric vehicles as a new power source for electric utilities. Transp Res Part D Transp Environ 2(3):157–175

Kempton W, Tomić J (2005) Vehicle-to-grid power implementation: from stabilizing the grid to supporting large-scale renewable energy. J Power Sources 144(1):280–294

Klevmarken NA (1989) Panel studies: what can we learn from them? Eur Econ Rev 33(2):523–529

Kraft J, Kraft A (1978) Relationship between energy and GNP. J Energy Dev 3(2):401–403

Kuznets S (1955) Economic growth and income inequality. Am Econ Rev 45(1):1–28

Lanjouw JO, Mody A (1996) Innovation and the international diffusion of environmentally responsive technology. Res Policy 25(4):549–571. http://dx.doi.org/10.1016/0048-7333(95)00853-5

Lee CC, Chang CP (2008) Energy consumption and economic growth in Asian economies: a more comprehensive analysis using panel data. Resour Energy Econ 30(1):50–65. http://dx.doi.org/10.1016/j.reseneeco.2007.03.003

Lehner B, Czisch G, Vassolo S (2005) The impact of global change on the hydropower potential of Europe: a model-based analysis. Energy Policy 33(7):839–855

Li J, Colombier M (2009) Managing carbon emissions in China through building energy efficiency. J Environ Manag 90(8):2436–2447. http://dx.doi.org/10.1016/j.jenvman.2008.12.015

Lipp J (2007) Lessons for effective renewable electricity policy from Denmark, Germany and the United Kingdom. Energy Policy 35(11):5481–5495

Luzzati T, Orsini M (2009) Investigating the energy-environmental Kuznets curve. Energy 34(3):291–300. http://dx.doi.org/10.1016/j.energy.2008.07.006

Magnani E (2000) The Environmental Kuznets Curve, environmental protection policy and income distribution. Ecol Econ 32(3):431–443. http://dx.doi.org/10.1016/S0921-8009(99)00115-9

Martinot E, Sawin J (2012) Renewables global status report. Renewables 2012 Global Status Report, REN21. http://www.martinot.info/REN21_GSR2012.pdf

Millimet DL, List JA, Stengos T (2003) The environmental Kuznets curve: real progress or misspecified models? Rev Econ Stat 85(4):1038–1047

Moomaw WR, Unruh GC (1997) Are environmental Kuznets curves misleading us? The case of CO2 emissions. Environ Dev Econ 2:451–463

Mundlak Y (1978) On the pooling of time series and cross section data. Econometrica 46(1):69–85

Narayan PK, Prasad A (2008) Electricity consumption–real GDP causality nexus: evidence from a bootstrapped causality test for 30 OECD countries. Energy Policy 36(2):910–918

Nordhaus WD (2002) Modeling induced innovation in climate-change policy. In: Nordhaus WD (ed) Arunf Grobler, Nebojsa Nakicenovic, Technological change and the environment. Resources for the Future, Washington, DC, pp 182–209

Oberthür S, Roche Kelly C (2008) EU leadership in international climate policy: achievements and challenges. Int Spectator 43(3):35–50

Panayotou T (1993) Empirical tests and policy analysis of environmental degradation at different stages of economic development, World Employment Programme Research, WEP Working Paper 2-22/WP.238, International Labour Office, Geneva

Panayotou T (1997) Demystifying the environmental Kuznets curve: turning a black box into a policy tool. Environ Dev Econ 2(4):465–484

Pearce D (1991) The role of carbon taxes in adjusting to global warming. Econ J 101(407):938–948

Popp DC (2001) The effect of new technology on energy consumption. Resour Energy Econ 23(3):215–239

Popp D (2003) Pollution control innovations and the Clean Air Act of 1990. J Policy Anal Manage 22(4):641–660

Popp D (2005) Lessons from patents: using patents to measure technological change in environmental models. Ecol Econ 54(2–3):209–226. http://dx.doi.org/10.1016/j.ecolecon.2005.01.001

Popp D (2006) International innovation and diffusion of air pollution control technologies: the effects of NOX and SO2 regulation in the US, Japan, and Germany. J Environ Econ Manag 51(1):46–71. http://dx.doi.org/10.1016/j.jeem.2005.04.006

Popp D, Hascic I, Medhi N (2011) Technology and the diffusion of renewable energy. Energy Economics 33(4):648–662. http://dx.doi.org/10.1016/j.eneco.2010.08.007

Prais SJ, Winsten CB (1954) Trend estimators and serial correlation, Cowles Commission Discussion Paper No. 383 (Chicago)

Premalatha M, Tauseef S, Abbasi T, Abbasi S (2013) The promise and the performance of the world's first two zero carbon eco-cities. Renew Sust Energ Rev 25:660–669

Pudjianto D, Ramsay C, Strbac G (2007) Virtual power plant and system integration of distributed energy resources. Renew Power Gener IET 1(1):10–16

Romm J (2002) The internet and the new energy economy. Resour Conserv Recycl 36(3):197–210

Ruiz N, Cobelo I, Oyarzabal J (2009) A direct load control model for virtual power plant management. Power Syst IEEE Trans 24(2):959–966

Saner D, Juraske R, Kübert M, Blum P, Hellweg S, Bayer P (2010) Is it only CO2 that matters? A life cycle perspective on shallow geothermal systems. Renew Sust Energ Rev 14(7):1798–1813

Schleisner L (2000) Life cycle assessment of a wind farm and related externalities. Renew Energy 20(3):279–288

Selden TM, Song D (1994) Environmental quality and development: is there a Kuznets curve for air pollution emissions? J Environ Econ Manag 27(2):147–162. http://dx.doi.org/10.1006/jeem.1994.1031

Shafik N (1994) Economic development and environmental quality: an econometric analysis. Oxford Econ Pap 46:757–773

Shafik N, Bandyopadhyay S (1992) Economic growth and environmental quality: time series and cross country evidence, vol 904. World Bank-free PDF. http://econ.worldbank.org/external/default/main?pagePK=64165259&theSitePK=469072&piPK=64165421&menuPK=64166322&entityID=000009265_3961003013329

Sinha A (1993) Modelling the economics of combined wind/hydro/diesel power systems. Energy Convers Manag 34(7):577–585

Soytas U, Sari R (2009) Energy consumption, economic growth, and carbon emissions: challenges faced by an EU candidate member. Ecol Econ 68(6):1667–1675

Stata Quick Reference and Index (2012) Stata Press, College Station, Texas

Stern DI (1998) Progress on the environmental Kuznets curve? Environ Dev Econ 3(2):173–196

Stern DI (2000) A multivariate cointegration analysis of the role of energy in the US macroeconomy. Energy Economics 22(2):267–283

Stern DI (2002) Explaining changes in global sulfur emissions: an econometric decomposition approach. Ecol Econ 42(1–2):201–220. http://dx.doi.org/10.1016/S0921-8009(02)00050-2

Stern DI, Common MS, Barbier EB (1996) Economic growth and environmental degradation: the environmental Kuznets curve and sustainable development. World Dev 24(7):1151–1160. http://dx.doi.org/10.1016/0305-750X(96)00032-0

Suri V, Chapman D (1998) Economic growth, trade and energy: implications for the environmental Kuznets curve. Ecol Econ 25(2):195–208

Torras M, Boyce JK (1998) Income, inequality, and pollution: a reassessment of the environmental Kuznets Curve. Ecol Econ 25(2):147–160. http://dx.doi.org/10.1016/S0921-8009(97)00177-8

Wagner M (2008) The carbon Kuznets curve: a cloudy picture emitted by bad econometrics? Resour Energy Econ 30(3):388–408

Williges K, Lilliestam J, Patt A (2010) Making concentrated solar power competitive with coal: the costs of a European feed-in tariff. Energy Policy 38(6):3089–3097. doi:10.1016/j.enpol.2010.01.049

Wolde-Rufael Y (2005) Energy demand and economic growth: the African experience. J Policy Model 27(8):891–903

Wooldridge JM (2002) Econometric analysis of cross section and panel data. The MIT Press, Cambridge, MA

Yu ES, Hwang BK (1984) The relationship between energy and GNP: further results. Energy Economics 6(3):186–190

Zhang XP, Cheng XM (2009) Energy consumption, carbon emissions, and economic growth in China. Ecol Econ 68(10):2706–2712

Chapter 8
Summary and Conclusion

Ongoing concerns about climate change have made renewable energy sources an important component of the world energy consumption portfolio. Renewable energy technologies could reduce CO_2 emissions by replacing fossil fuels in the power generation industry and the transportation sector. Because of some negative and irreversible externalities in conventional energy production, it is necessary to develop and promote renewable energy supply technologies and demand for renewable energy. Power generation using renewable energy sources should be increased in order to decrease the unit cost of generation. Energy consumption depends on several factors including economic progress, population, energy prices, weather, and technology.

Although renewable energy has been used in the past as a major source of energy for many centuries, it currently constitutes only a small percentage of the world's total primary energy supply. Based on an IEA report, renewable energy subsidies sharply increased to 88 billion dollars in 2011, which shows a growth rate of 24 % from 2010. According to *GSR 2012* (Martinot and Sawin 2012), new investments in renewable energy sources worldwide increased to 257 billion dollars in 2011, which is twice the amount of investment in 2007 and six times higher than the investment level in 2004. Hydro, wind, and solar energy are the main sources of renewable energy in many countries.

8.1 Economic Growth and Energy Consumption Relationship

The level of economic activities plays a key role in energy consumption, which is considered a key driver of the functioning and development of energy markets. Many scholars have studied the relation between energy consumption and economic growth. Generally, four hypotheses are defined in this regard. The first hypothesis

© Springer Science+Business Media Singapore 2015
A. Heshmati et al., *The Development of Renewable Energy Sources and its Significance for the Environment*, DOI 10.1007/978-981-287-462-7_8

called the neutral hypothesis suggests that if there is no causality, then the energy consumption is not related to GDP. Therefore, neither conservative nor expansive policies affect economic growth. The second hypothesis concerns unidirectional causality, which may be from economic growth to energy consumption (conservation hypothesis) or energy consumption to economic growth (growth hypothesis), forming the basis for the third hypothesis. The fourth feedback hypothesis is applicable if there is bidirectional causality. Depending on the hypothesis, energy policies have different effects on economic growth.

Early research on this causality topic was published in the 1970s, and an extensive body of literature has evolved since then. Many researchers have studied the causality relationship between energy consumption and economic growth using different variables. Using various approaches and methodologies, previous research studied individual countries and groups of countries to investigate the effect of governmental energy policy on economic growth.

As Table 2.4 showed, some countries showed different results within the same period regardless of the methodology used. It should also be noted that this analysis considers individual relationships between two variables (here, energy consumption and economic growth). Therefore, the results need to be interpreted with caution when used as a basis for making decisions regarding energy policy. Other parameters such as technological innovation and governmental taxes may affect this relationship. The effects of energy policies depend on conditions in the country, the methodology used, and the effects of time (traditionally representing technological advancement) in the data sample. The effects of interactions with other variables should also be considered.

Although many researchers have studied the relationship between energy consumption and economic growth, they have rarely focused on causal relationships in renewable energy consumption. Renewable energy technologies could enable countries with good solar or wind resources to deploy these energy sources to meet their domestic demand. Two main solutions may be implemented to reduce CO_2 emissions and overcome the problem of climate change: (i) replacing fossil fuels with renewable energy sources as much as possible and (ii) enhancing energy efficiency. Energy efficiency in an electricity network could be considered in different stages, including power generation (here, the focus is on small amounts), transmission, distribution, and consumption.

8.2 Energy Efficiency Technologies

Improved energy efficiency is an important way to reduce energy use and, thereby, CO_2 emissions and to overcome the climate change problem. Energy efficiency for electricity networks can be considered in different stages, such as power generation, transmission, distribution, and consumption. For this purpose, different energy efficiency technologies are available, including electric vehicles, combined heat and power, virtual power plants, and smart grids. Each of these technologies has been discussed in detail and their performances compared. The improved flexibility of the smart grid network will help develop different renewable energy sources, such as solar and wind power. The smart grid enables households to have their own devices to produce renewable energy and trade surpluses of electricity. Furthermore, households could manage energy consumption using intelligent control systems. Households that generate electricity using their own resources may enter their surplus electricity in the network and withdraw it at other times when needed. This power exchange between the distributed generation owner and the network requires appropriate management.

Electric vehicles have the potential to be used for both power generation and storage. Given the fact that transportation is a main contributor to the problem of emissions, improving fuel efficiency with the adoption of electric vehicle technology on a large scale will enable greater energy savings and CO_2 emission reduction. Advances in the smart grid technology impact the large-scale use of electric vehicles and enhance the efficiency of the technology. However, managing load and supply fluctuations is a challenge. Combined heat and power technologies provide substantial gains in efficiency. The technology offers efficient use of fuel by preventing the discarding of energy as waste heat. A significant part of heat can be transformed into a by-product for heating buildings, adding to its economic value and improving energy efficiency. A virtual power plant is a cluster of distributed energy resources controlled and managed by a central control unit, allowing for the possibility to control home appliances to optimize load reductions. It helps combat the energy waste problem due to distance and transmission losses.

The driving force for using renewable energy technologies are energy security, economic impacts, and CO_2 emission reduction. The level of insecurity is reflected by the risk of supply disruption and the estimated costs of security itself. The emphases for the economic impacts are job creation, industrial innovation, and balance of payment. Renewable energy technologies could enable countries with good solar or wind resources to implement these energy sources to meet their own domestic demand. Moreover, the cost of importing fuels can affect economic growth. If these countries could reduce their balance of payment by producing their own renewable energy to replace their dependence on fossil fuels, they could expand their capacity for investment in other sectors. Renewable energy technologies could reduce CO_2 emissions by replacing fossil fuels in the power generation industry and transportation sector. Life-cycle CO_2 emission for renewable energy technologies is expected to be lower than fossil fuels.

This review of renewable energy generation and efficiency technologies has provided detailed and useful information that can be used in the decision-making of different stakeholders in the rapidly developing energy market. Each technology has both advantages and disadvantages that vary by location, availability, the technological capability of producers, financial limitations, and environmental considerations. Each municipality, region or country has different initial conditions that determine the energy mix that can be produced at the lowest cost while minimizing the harm done to the environment. Thus, there is no single solution to every energy need and problem but rather an optimal location-specific solution among a set of possible renewable solutions.

8.3 Selected Technologies of Renewable-Based Power Generation

Power generation using renewable energy sources can be used to decrease the unit cost of energy and to make them a competitive alternative to the conventional energy sources. Two main solutions may be implemented to reduce CO_2 emissions and to overcome the problem of climate change: replacing fossil fuels with renewable energy sources as much as possible and enhancing energy efficiency regardless of type. In this review, we considered hydro, wind, solar, and geothermal sources because of their significant contribution to power generated by renewable sources. In the first part of this review, we discussed state-of-the-art methods for technical and economic feasibility in the implementation of renewable energy sources as well as the possibility of their combined use and substitution.

Renewable energy production and supply is continuously increasing on a global level. Following the drastic increase in oil price and its impacts on coal and gas prices, a large amount of investment has been made over recent years in renewable energy. These advancements in technology have enabled countries to produce renewable energy in larger quantities and more cost-effectively. Due to negative externalities associated with conventional energy extraction and consumption, it is necessary to promote and develop renewable energy supply and consumption. Renewable energy sources act as substitutes for fossil fuels and reduce emissions. In the short term, some renewable technologies may not be comparable to conventional fuels in the scope of production costs and transmission, but they could be comparable if we consider their associated positive externalities, such as their environmental and social effects. Also, it should be noted that economies of scale could play a key role in reducing the unit cost of production.

Transmission and distribution costs and technologies do not differ much among the conventional and renewable energy sources. In this review, we have presented detailed facts about the main renewable energy supply technology developments, including hydro, wind, solar, and geothermal, and other sources such as biomass, ocean waves, and tides in brevity. The emphasis has been on current production capacity and the estimated capacity as well as development costs, which are sunk. We have also presented empirical findings from comparative studies of alternative energy technologies.

Hydropower is the largest renewable energy source for power generation around the world. Despite its large energy generation contribution, its development is difficult due to a high initial fixed investment cost and environmental and population relocation costs. Hydropower is attractive due to a combined supply of water for agriculture, household, recreation, and industrial use. Additionally, it can store water and energy that can be used for both base and peak load power generations. The availability of funding, political and market risks, resource allocation priorities, and local environmental concerns are considered to be barriers to the development of hydropower capacity.

The installed wind power capacity has also been increasing, especially in countries like China, the USA, Germany, and Denmark. Advantages of wind power plants include the installation as turnkey contracts within a short period, a lower investment compared to nuclear and hydroelectric plants, economies of mass production, an absence of fuel costs, and low operation and maintenance costs. The problems associated with the use of wind power include intermittency of wind energy and an added cost for power transmission to users. Generation cost is dependent on location, feasibility, and the minimum required speed for wind turbines.

China has developed its own solar power capacity, decreasing the cost of generation due to the availability of cheap labor and public subsidies. Another source of the reduced costs is advances in and the high efficiency in concentrated solar power technologies in the USA. The negative effects include land, material, and chemical use and impacts on buildings' aesthetics. The performance is dependent on location.

Geothermal energy has been used throughout history for bathing, heating, and cooking. The geothermal gradient and permeability of rocks determine its economic implementation feasibility. Unlike wind and solar power, geothermal energy is continuously available through the year, although the technology has some negative environmental effects.

8.4   The Main Support Mechanisms to Finance Renewable Energy

Environmental regulations could affect the commercial policies and strategic decisions made by companies and users. The interactions between players and organizations (state and nonstate) in markets and their reciprocal influence are important in making and implementing an effective policy. Regarding renewable energies in addition to parameters such as decreasing technology costs because of advancement and economies of scale, the rapid growth of renewable energy was achieved mainly by support policies. As some sources of energy (e.g., natural gas and coal) are available in the market at low prices, renewable energy would not be economical without active government support. As our findings indicated, market regulation, technological innovation, and environmental tax play significant roles in meeting targets.

Alternative policies for environmental protection are applied in the form of economics (incentive) or regulations (nonincentive). Economic policies could be an incentive for using renewable energy or charging taxes imposed on emission generation or fossil fuel consumption. Three types of support mechanisms are widely employed: (i) feed-in tariffs, (ii) tax incentives, and (iii) tradable green certificates.

Tax credits could be applied to the investment, production, or consumption segments of electricity generated by renewable energy sources. The main reason that this instrument is attractive is that it makes cash available. Therefore, it could be an important financial incentive for private investors as well as an opportunity to make small investments because it directly increases investor liquidity. Couture and Gagnon (2010) pointed to three essential provisions for the success of FIT policies: guaranteed access to the grid, stable and long-term power purchase agreements, and calculation of prices on the basis of unit costs of power generated by renewable energy sources.

In contrast to the FIT policy, which is price based (fixed price and premium price), the RPS policy is quantity based. This instrument requires companies to increase the amount of power generated by renewable energy sources. The comparison of FIT and RPS policies showed that the FIT policy is preferred when a policy to develop renewable energy sources with a low level of risk for investors is required. However, the RPS policy is appropriate when a market-view policy is applied by the government. Although Europe has attempted to organize a single harmonized FIT system, it is believed to be impossible due to the vast differences in policies across the member countries. The RPS system has not been developed in Europe as most European countries have previously instituted the FIT system. In this regard, it seems that FIT polices are suitable to encourage the development of renewable energy sources. However, the RPS system should be applied after the implementation of renewable energy sources has reached a certain level. Considering technological progress and the cost reduction for power generation by renewable energy sources, we suggest that support mechanism policies should be reconsidered from the financial point of view.

8.5 Market Design for Trading Commoditized Renewable Energy

Energy efficiency is one of the most important approaches to reducing electricity consumption in current times as well as in the future. ICT plays an important role in achieving this target. In particular, a smart grid can achieve a higher rate of energy efficiency by integrating IT and establishing interactions between suppliers and customers. The concept of a grid market is already created for computing resources. For instance, some scholars have already expressed views about the potential of virtual power plants and micro sources.

Considering the outlook of the renewable energy market, a marketplace where small volumes of generated electricity can be traded is needed. Without marketplace development, renewable energy usage is limited to only covering individual demand. However, if it is connected to a commercial marketplace, its progress could be limitless, similar to the wholesale electricity market. Here, we have tried to design a similar market for renewable energy resources. At first, we need to recognize the components of a smart network. All components should have this capability to be managed by an automated system. We should keep in mind that a new technology to deploy smart control and management is required. Also, it should be compatible with the current system. Next, we analyzed the regulatory framework and introduced some instruments for the efficient operation of the proposed market.

Our proposed marketplace is a place to trade electricity power that is produced by households. In fact, there should be a balance between supply and demand in order to prevent power blockages. The microgrid concept could be considered here. Many households could be connected to each other as micro generators, and then, they would be able to reduce energy consumption at peak periods and cover their demand at other times. Therefore, through marketplace development and an increasing number of households in the market, the demand for increasing the capacity of new power generation to cover electricity consumption could be reduced. The primary infrastructure required to establish this market includes the following: communication, science–technology infrastructure, storage facilities, and device provisioning.

By analyzing the incentives of customers to participate in a smart network and by studying the structure of distributed energy resources in order to establish a market, we have identified the existence of five potential major restrictions to the wide use of renewable energy. The analysis of these restrictions has indicated that a solution can be a market for trading renewable energy generated by micro sources. This market has the capacity to provide support and consultancy services in order to help customers integrate demand response programs into their existing IT infrastructures.

The proposed marketplace consists of several players, such as major power generators (including fossil fuel and non-fossil fuel generators), distributors, service companies, and consumers (including industries, commercial buildings, households, and farms). Therefore, interaction between these parties is an important factor. This interaction should be considered in discussions about regulations and public incentives and supports. Furthermore, external factors should be taken into account in order to maintain stability in our policy and avoid uncertainty in market players regarding policies and regulations. In order to establish a successful market-based microgrid, we should use a combination of instruments, including command-control, self-regulation, education and information, and economic instruments as a regulatory framework. Governmental regulations should be set to monitor, control, and manage the market efficiently.

Considering that the market mechanism is the core of a marketplace, we defined different parameters such as the start time, unit of duration, and unit of trade for the optimal functioning of the proposed renewable marketplace. We explained our proposals for the smooth functioning of the market mechanism. We suggested some indices to evaluate the market's performance. These indices include sales revenue, trading volume, storage capacity, installed generators, and CO_2 emission reduction. Finally, we proposed a combination of a few complementary policies for building the necessary infrastructure and removing barriers and the financial support for the creation, development, and effective operation of markets. Furthermore, we emphasized that the stability and sustainability of any effective policy and regulation is one of the most important factors to enable the success of the new market operation.

8.6   The Climate Impacts of Alternative Energy Technologies

During the last three decades, two different approaches have been applied in the context of natural resources. The first approach considers the effects of natural resources on economic growth, and the second approach explains the effects of economic growth on pollution, such as CO_2 emissions. Many researchers have studied the relationship between energy consumption and economic growth. The second approach takes into account the environmental effects of economic growth. Following an empirical study by Grossman and Krueger (1991), many scholars analyzed the relation between economic growth and environmental pollution. The environmental consequences of economic development are crucial in any discussion about sustainable economic growth.

We evaluated the impact of alternative energy production, technological innovation for mitigation or adaptation to climate change, and market regulation of CO_2 emission reduction. Our model estimated the effectiveness of these parameters for a panel of 15 countries in the EU-15 from 1995 to 2010. We selected these countries because they are in the forefront of renewable energy development and applied market regulation to mitigate climate change. The earliest EKC function was made as a simple quadratic function of the levels of income. We employed additional explanatory variables in order to evaluate environmental effects. The number of patents related to energy technologies and ICT per million inhabitants was applied in the model instead of R&D expenditure in order to measure the effects of technological innovation on CO_2 emission reduction. In addition, environmental tax per capita was considered to show that market regulation influences CO_2 emissions.

Usually, the model described above has been estimated using panel data. Most studies estimated both fixed-effect and simple random-effect models. Our estimation method differed from most studies because it used feasible generalized least squares (FGLS) and correct heteroscedasticity and autocorrelation. Although heteroscedasticity and autocorrelation in residuals are serious problems, none of the early studies reported diagnostic tests of models. Consistent with Harbaugh et al. (2002), Millimet et al. (2003), Stern (2004), and Wagner (2008), our findings confirmed that an inverted U-shaped relationship between CO_2 emissions and GDP obtained with commonly used methods is entirely spurious because of several major econometric problems.

The relation between CO_2 emissions and GDP showed a concave curve for the EU-15 countries, which contrasts many studies that focused on the EKC and claimed an inverted U-shaped curve relationship between emissions per capita and GDP per capita in developed countries. In addition, the results showed a convex formation between CO_2 emissions per capita with the number of ICT patents per million inhabitants and technological change over time represented by a time trend. Therefore, we found a decrease in the increasing rate of CO_2 emissions regarding ICT patents and technological change over time. Furthermore, there was a convex formation between CO_2 emissions per capita in relation to energy patent applications per million inhabitants and environmental tax per capita. The results showed an increase in the decreasing rate of CO_2 emissions in relation to energy patents and environmental tax. Based on our findings, the policies developed for environmental tax, energy patent applications, and renewable energy generation could mitigate CO_2 emissions.

The effects of interactions among policies play a key role in reducing CO_2 emissions. Appropriate market instruments should be selected to mitigate CO_2 emissions. These instruments could be complementary or sequential. We recommend avoiding policies that affect each other negatively. We cannot assume that increasing GDP per capita means that more will be spent on environment quality because the latter is based on consumer behavior. Our results showed that environmental tax per capita has a negative impact on CO_2 emissions, but the elasticity of environmental tax in relation to CO_2 emissions is positive. Tax rates should be high enough to make a sensitive reduction in CO_2 emissions. A reduction in CO_2 emissions may not be achieved unless the related elasticities are known with acceptable certainty.

The results of this research could be important even in developing countries. They are able to achieve the same level of environmental quality before achieving high levels of GDP per capita if appropriate policies are developed by their governments. Countries that are attempting to develop ecocities, such as Dongtan City in China and Masdar City in the UAE, may find our findings useful. We have analyzed a wide range of strategies used in different countries, which has indicated opportunities for the UAE government to formulate policies to promote renewable energy in Masdar City. An advantage of Masdar City that distinguishes it from other ecocities is that it has the potential to establish the first market for trading renewable energy. The micro certificate, a type of partial trading in the RPS mechanism, could be applied in this market.

References

Couture T, Gagnon Y (2010) An analysis of feed-in tariff remuneration models: implications for renewable energy investment. Energy Policy 38(2):955–965. doi:10.1016/j.enpol.2009.10.047

Grossman GM, Krueger AB (1991) Environmental impacts of a North American free trade agreement, National Bureau of Economic Research, NBER Working Paper No. 3914

Harbaugh WT, Levinson A, Wilson DM (2002) Reexamining the empirical evidence for an environmental Kuznets curve. Rev Econ Stat 84(3):541–551

Martinot E, Sawin J (2012) Renewables global status report. Renewables 2012 Global Status Report, REN21. http://www.martinot.info/REN21_GSR2012.pdf

Millimet DL, List JA, Stengos T (2003) The environmental Kuznets curve: real progress or misspecified models? Rev Econ Stat 85(4):1038–1047

Stern DI (2004) The rise and fall of the environmental Kuznets curve. World Dev 32(8):1419–1439

Wagner M (2008) The carbon Kuznets curve: a cloudy picture emitted by bad econometrics? Resour Energy Econ 30(3):388–408

Appendix A

Table A.1 Feasible generalized least squares estimation result for alternative model with environmental tax/fossil fuel consumption

Variables	Model 1B		Model 2B		Model 3B	
	Coeff.	P>|z|	Coeff.	P>|z|	Coeff.	P>|z|
LnGDP/P	−1.148	0.000	−1.356	0.000	−1.389	0.000
(LnGDP/P)^2	0.208	0.000	0.285	0.000	0.573	0.000
LnRen/TPG			−0.002	0.455	−0.123	0.084
(LnRen/TPG)^2			0.028	0.000	0.027	0.000
LnPateng/P			−0.004	0.000	−0.094	0.004
(LnPateng/P)^2			−0.004	0.000	−0.001	0.495
LnPatict/P			0.007	0.000	0.212	0.000
(LnPatict/P)^2			−0.004	0.000	0.001	0.482
LnEvt/egp			−0.911	0.000	−0.175	0.022
(LnEvt/egp)^2			0.709	0.000	0.076	0.000
Trend			0.033	0.000	0.052	0.000
Trend^2			−0.002	0.000	−0.002	0.000
(LnGDP/P)(LnRen/TPG)					0.124	0.000
(LnGDP/P)(LnPateng/P)					0.000	0.967
(LnGDP/P)(LnPatict/P)					−0.045	0.000
(LnGDP/P)(LnEvt/egp)					−0.273	0.000
(LnGDP/P)Trend					−0.019	0.000
(LnRen/TPG)(LnPateng/P)					−0.018	0.000
(LnRen/TPG)(LnPatict/P)					0.011	0.001
(LnRen/TPG)(LnEvt/egp)					−0.065	0.000
(LnRen/TPG)Trend					−0.005	0.000

(continued)

© Springer Science+Business Media Singapore 2015
A. Heshmati et al., *The Development of Renewable Energy Sources and its Significance for the Environment*, DOI 10.1007/978-981-287-462-7

Table A.1 (continued)

Variables	Model 1B Coeff.	P>|z|	Model 2B Coeff.	P>|z|	Model 3B Coeff.	P>|z|
(LnPateng/P)(LnPatict/P)					0.001	0.750
(LnPateng/P)(LnEvt/egp)					0.012	0.033
(LnPateng/P)Trend					−0.001	0.190
(LnPatict/P)(LnEvt/egp)					−0.009	0.306
(LnPatict/P)Trend					−0.001	0.048
(LnEvt/egp)Trend					0.007	0.000
Const.	3.672	0.000	6.145	0.000	3.704	0.000
Number of obs	240		240		240	
Number of groups	15		15		15	
Time periods	16		16		16	
Wald chi2	516.71		56,653.30		8,904.97	
Prob > chi2	0.0000		0.0000		0.0000	

Table A.2 CO_2 elasticities in EU-15 countries over 1995–2010

	GDP	Regenp	Ptgcap	Pticap	Evtegp	Trend
1995	0.6735	−0.1331	0.0081	−0.0077	−0.1692	0.0306
1996	0.6453	−0.1377	0.0073	−0.0083	−0.1619	0.0263
1997	0.5490	−0.1443	0.0041	−0.0036	−0.1482	0.0224
1998	0.5368	−0.1424	0.0025	−0.0041	−0.1476	0.0178
1999	0.4885	−0.1504	0.0028	−0.0049	−0.1348	0.0139
2000	0.3971	−0.1594	−0.0011	0.0007	−0.1235	0.0101
2001	0.3559	−0.1672	−0.0022	0.0011	−0.1140	0.0060
2002	0.3621	−0.1806	0.0017	−0.0057	−0.0976	0.0026
2003	0.5013	−0.1742	0.0027	−0.0163	−0.1132	−0.0039
2004	0.5900	−0.1720	0.0018	−0.0227	−0.1210	−0.0098
2005	0.5811	−0.1791	0.0005	−0.0241	−0.1132	−0.0144
2006	0.6034	−0.1839	−0.0022	−0.0256	−0.1121	−0.0198
2007	0.6895	−0.1840	−0.0024	−0.0325	−0.1174	−0.0253
2008	0.7206	−0.1841	−0.0046	−0.0348	−0.1191	−0.0306
2009	0.5946	−0.1974	−0.0076	−0.0293	−0.0918	−0.0331
2010	0.5906	−0.1898	−0.0083	−0.0301	−0.0919	−0.0356
Mean	0.5550	−0.1675	0.0002	−0.0155	−0.1235	−0.0027
Std dev	0.1108	0.0203	0.0047	0.0127	0.0232	0.0221

Table A.3 CO_2 elasticities in EU-15 by country

	GDP	Regenp	Ptgcap	Pticap	Evtegp	Trend
Austria	0.8248	−0.0763	−0.0290	0.0008	−0.2518	−0.0133
Belgium	0.5390	−0.2054	0.0208	−0.0277	−0.1259	0.0005
Germany	−0.0152	−0.3248	0.0314	−0.0388	0.1999	0.0108
Denmark	0.6725	−0.2001	0.0050	−0.0324	−0.0759	−0.0051
Spain	0.4338	−0.1578	−0.0101	0.0030	−0.1308	0.0011
Finland	0.7189	−0.0883	−0.0180	−0.0011	−0.2535	−0.0122
France	0.6031	−0.1479	−0.0036	−0.0099	−0.1779	−0.0053
Greece	0.1263	−0.2330	0.0046	0.0003	−0.0021	0.0114
Ireland	0.6817	−0.1781	0.0172	−0.0361	−0.1221	−0.0010
Italy	0.4322	−0.1737	−0.0031	−0.0085	−0.0891	0.0003
Luxembourg	1.5487	−0.0399	−0.0150	−0.0412	−0.3839	−0.0223
Netherlands	0.4844	−0.2082	0.0171	−0.0289	−0.0796	−0.0010
Portugal	0.1175	−0.1941	−0.0146	0.0164	−0.0217	0.0093
Sweden	0.8623	−0.0703	−0.0253	−0.0024	−0.2781	−0.0154
UK	0.2947	−0.2141	0.0254	−0.0260	−0.0606	0.0020
Mean	0.5550	−0.1675	0.0002	−0.0155	−0.1235	−0.0027
Std dev	0.3795	0.0742	0.0190	0.0183	0.1381	0.0098

Appendix B

Table B.1 Feasible generalized least squares estimation result for alternative models with normalized variables

Variables	Model 1A Coeff.	P>|z|	Model 2A Coeff.	P>|z|	Model 3A Coeff.	P>|z|
LnGDP/P	−1.756	0.000	−3.080	0.000	−0.500	0.328
$(LnGDP/P)^2$	1.101	0.000	1.742	0.000	1.535	0.000
LnRen/TPG			0.018	0.102	0.222	0.001
$(LnRen/TPG)^2$			0.033	0.000	0.072	0.000
LnPateng/P			−0.002	0.000	−0.071	0.000
$(LnPateng/P)^2$			0.000	0.003	0.000	0.191
LnPatict/P			0.034	0.000	1.404	0.000
$(LnPatict/P)_2$			−0.022	0.000	−0.005	0.409
LnEvt/P			−0.002	0.000	0.037	0.000
$(LnEvt/P)^2$			0.000	0.000	0.000	0.000
Trend			0.134	0.000	0.694	0.000
$Trend^2$			−0.079	0.000	−0.061	0.000
(LnGDP/P)(LnRen/TPG)					−0.286	0.000
(LnGDP/P)(LnPateng/P)					0.061	0.000
(LnGDP/P)(LnPatict/P)					−1.339	0.000
(LnGDP/P)(LnEvt/P)					−0.026	0.000
(LnGDP/P)Trend					−0.616	0.000
(LnRen/TPG)(LnPateng/P)					0.003	0.039
(LnRen/TPG)(LnPatict/P)					−0.011	0.262
(LnRen/TPG)(LnEvt/P)					−0.001	0.389
(LnRen/TPG)Trend					0.065	0.000

(continued)

© Springer Science+Business Media Singapore 2015
A. Heshmati et al., *The Development of Renewable Energy Sources and its Significance for the Environment*, DOI 10.1007/978-981-287-462-7

Table B.1 (continued)

Variables	Model 1A Coeff.	P>|z|	Model 2A Coeff.	P>|z|	Model 3A Coeff.	P>|z|
(LnPateng/P)(LnPatict/P)					0.002	0.255
(LnPateng/P)(LnEvt/P)					0.001	0.000
(LnPateng/P)Trend					0.001	0.231
(LnPatict/P)(LnEvt/P)					−0.012	0.000
(LnPatict/P)Trend					−0.027	0.002
(LnEvt/P)Trend					−0.003	0.000
Const.	1.620	0.000	2.231	0.000	−0.135	0.625
Number of obs	240		240		240	
Number of groups	15		15		15	
Time periods	16		16		16	
Wald χ^2	516.71		1,389.32		2,549.09	
Prob > χ^2	0.0000		0.0000		0.0000	

Table B.2 CO_2 elasticities in EU-15 countries over 1995–2010

	GDP	Regenp	Ptgcap	Pticap	Evtcap	Trend
1995	0.9436	0.0947	−0.0030	0.0013	−0.0015	0.1066
1996	0.8367	0.1020	−0.0036	0.0104	−0.0021	0.0941
1997	0.5785	0.1086	−0.0040	0.0216	−0.0030	0.0843
1998	0.4677	0.1087	−0.0042	0.0256	−0.0040	0.0669
1999	0.2950	0.1206	−0.0043	0.0290	−0.0051	0.0549
2000	0.0667	0.1311	−0.0041	0.0296	−0.0055	0.0458
2001	−0.0566	0.1418	−0.0042	0.0326	−0.0060	0.0344
2002	−0.0298	0.1604	−0.0041	0.0289	−0.0061	0.0232
2003	0.1597	0.1542	−0.0041	0.0113	−0.0066	−0.0092
2004	0.3091	0.1491	−0.0045	0.0034	−0.0066	−0.0357
2005	0.2750	0.1546	−0.0047	0.0014	−0.0068	−0.0521
2006	0.3106	0.1549	−0.0049	−0.0059	−0.0066	−0.0731
2007	0.4647	0.1529	−0.0048	−0.0224	−0.0065	−0.0995
2008	0.4970	0.1499	−0.0048	−0.0302	−0.0066	−0.1221
2009	0.3197	0.1593	−0.0053	−0.0145	−0.0064	−0.1256
2010	0.3912	0.1605	−0.0050	−0.0086	−0.0056	−0.1335
Mean	0.3643	0.1377	−0.0044	0.0071	−0.0053	−0.0088
Std dev	0.2741	0.0231	0.0006	0.0199	0.0017	0.0838

Table B.3 CO_2 elasticities in EU-15 by country

	GDP	Regenp	Ptgcap	Pticap	Evtcap	Trend
Austria	0.4872	0.0156	−0.0031	−0.0504	−0.0067	−0.0855
Belgium	−0.1780	0.2515	0.0033	−0.0941	−0.0095	0.0158
Germany	1.2717	0.2027	−0.0352	0.5266	0.0066	0.1459
Denmark	1.0774	0.1241	−0.0117	0.0838	−0.0016	−0.0119
Spain	0.2170	0.1300	0.0028	−0.0890	−0.0038	−0.0123
Finland	−0.1210	0.0792	−0.0048	0.0065	−0.0104	−0.0477
France	−0.1013	0.1436	0.0021	−0.0974	−0.0095	−0.0347
Greece	0.3705	0.1995	0.0000	−0.0122	0.0011	0.0561
Ireland	0.3509	0.1761	0.0022	−0.0983	−0.0085	−0.0247
Italy	0.3745	0.1232	−0.0038	0.0073	−0.0039	−0.0053
Luxembourg	1.6103	0.0319	−0.0027	−0.1111	−0.0053	−0.1208
Netherlands	0.0669	0.2007	−0.0057	0.0365	−0.0081	0.0155
Portugal	0.6785	0.1185	−0.0050	0.0625	0.0046	0.0459
Sweden	0.2514	0.0362	−0.0049	−0.0174	−0.0090	−0.0769
UK	−0.8916	0.2329	0.0013	−0.0468	−0.0155	0.0087
Mean	0.3643	0.1377	−0.0044	0.0071	−0.0053	−0.0088
Std dev	0.6236	0.0738	0.0095	0.1564	0.0060	0.0645

Name Index

© Springer Science+Business Media Singapore 2015
A. Heshmati et al., *The Development of Renewable Energy Sources and its Significance for the Environment*, DOI 10.1007/978-981-287-462-7

Subject Index

© Springer Science+Business Media Singapore 2015
A. Heshmati et al., *The Development of Renewable Energy Sources and its
Significance for the Environment*, DOI 10.1007/978-981-287-462-7